STEM CELLS - LABORATORY AND CLINICAL RESEARCH

MORPHOGENETIC DEVELOPMENTAL PROGRAMS

STEM CELLS

STEM CELLS - LABORATORY AND CLINICAL RESEARCH

Additional books in this series can be found on Nova's website under the Series tab.

STEM CELLS - LABORATORY AND CLINICAL RESEARCH

MORPHOGENETIC DEVELOPMENTAL PROGRAMS

STEM CELLS

TATYANA BATYGINA

Nova Science Publishers, Inc.
New York

Library of Congress Cataloging-in-Publication Data

Batygina, T. B. (Tatiana Borisovna), 1927-
[Razrabotka teorii i vyiavlenie rezervov reproduktsii rastenii, morfogeneticheskie programmy i stvolovye kletki kak osnova ustoichivogo razvitiia. English]
Morphogenetic developmental programs, stem cells / Tatyana Batygina.
p. ; cm.
Includes bibliographical references and index.
ISBN 978-1-61209-576-9 (softcover : alk. paper)
1. Morphogenesis. 2. Stem cells. I. Title.
[DNLM: 1. Morphogenesis--physiology. 2. Stem Cells--physiology. QH 491]
QH491.B38313 2011
571'.833--dc22

2011002567

Published by Nova Science Publishers, Inc. † New York

In the memory of my dear parents
Zinaida Evdokimovna and Boris Mikhaylovich Gumensky

"Let the nature be your only goddess.
Treat her with boundless trust.
...Be faithful to her without being afraid
to lose your ambitions"
Auguste Rodin

Contents

Preface

Today's actual question is *creation of the science on stem cells* and development of its *theoretical bases*, taking into account comparative evolutionary cell biology and the *cardinal problem of the developmental biology – the problem of wholeness* for representatives of various kingdoms. This allows revealing correlations and studying correlative dependences of various structures at different levels of hierarchy.

Special attention should be paid at morphogenetic developmental programs and the reliability system.

Creation of any science requires a *certain preparation of mind* and is impossible without applying of *methodology* – one of the sciences studying human mind, the spirit in its highest manifestations and the technology of reasoning.

In current work the system of nontraditional ideas about the nature and role of stem cells in ontogenesis, reproduction and evolution of plants is proposed.

The *main properties of plant stem cells* have been developed by us, which has shown the *integrity* of morphogenous and reproductive processes at all stages of plant's life cycle. A special attention should be paid to the dynamics of the *stem cells' pool* and its correlative dependence with *niche*.

System approach and analysis of data on male and female generative structures in model objects, such as egg cell, double fertilization, plasmogamy and karyogamy, cell cycle, zygote, embryogenesis allowed us to *consider zygote* to be *the first stem cell.* Zygote is *a unique stem cell giving rise* not only for all *apical meristems*, but also for *stem cells of other orders*.

Zygote contains information about architectonics and development of organization of an individual, which is connected with gradients and correlations between its parts, and consequently with heterogeneity of different axes of a future organism.

The data of developmental biology and especially comparative embryology of representatives of various taxa of flowering plants enable to consider that *formation of stem cells derived from zygote, is probably typical for all organs – flower, stem, leaf, root* – and at all *stages of the life cycle – sporophyte, gametophyte*. Their localization is depending, first of all, on their functioning.

Not every cell possesses the totipotency and not every totipotent cell is able to become the stem cell, and not every stem cell can produce an individual.

One has to pay special attention to the role of somatic and stem cells in organization of embryo's body, in reproduction, and particularly, in changing of morphogenetic programs during ontogenesis and evolution. This view of the role of somatic cells also corresponds to the data of discoveries made in the middle of the XXth century: *stem cells, growing plants out of somatic cells, haploidy, parasexual hybridization, and embryoidogeny.*

I believe, in all these discoveries, *the somatic cell* to be *the corner stone* both for morphogenesis and ontogenesis, reproduction and evolution. It is probable that we are able to discuss *the evolution of somatic cells*.

A new phenomenon, embryoidogeny, was *for the first time* revealed by us in *1977 – a new type of vegetative propagation*, with an embryoid (somatic embryo) formed out of somatic cells being an elementary structural unit. Being, for the first time, studied by us from the position of stem cells and reproduction, this category is of great biological and practical importance (Batygina, 1977-2010).

Thus, *not only zygote*, but *somatic cell* as well is able to be the initial cell – stem cell of a plant, playing a significant role in ontogenesis and evolution.

The diversity of morphogenetic processes consists of *two components*: potentially *infinite activity of somatic cells* that is conditioned by the functioning of dormant centres' cells, being in intermediate differentiated state, and *loss of stemness – senescence and apoptosis*.

The plasticity and polyvariety of these processes are conditioned by the diversity of the morphogenesis and origin of vegetative and generative structures in which the peculiar *reserve of cells*, possessing *the stemness property*, is situated. Hence in plant morphogenetic processes even *the third component* is present – *restoring of stemness* in many respects depending on stemness degree.

The arsenal of the data received in the XX-XXI centuries on biology of development including stem cells biology has allowed us to propose and discuss *the three main perspective trends in the study of stem cells in life cycle of plants*:

- *gametes – zygote – embryo – individual – population – coenosis;*
- *different types of somatic cells considering their localization and function in the organism; possibility of homologization of some somatic cells with the egg cell, fertilized egg cell and zygote;*
- *apical meristem of shoot and root apices in which organization somatic cells take part.*

Integration of the three trends from the position of wholeness practically embraces the whole process of morphogenesis from the egg cell to the life cycle.

Progress in theoretical researches of *the new science about stem cells and their role in ontogenesis and evolution* will be connected with the *development of the theory of developmental biology, evolutionary developmental biology (Evo-Devo) and the reproduction theory.*

Chapter I

Introduction

The science of *individual development* and cognition of the essence and reasons of the morphogenic processes occurring in this period, regularities of origin and development of the first and further ontogenesis stages relives the band heyday, conditioned by the general progress of natural science and fall on the golden age of Russian biology (beginning of XX century).

The propagation occurs to be the main property of living matter and, as Charles Darwin proposed, the premise to natural selection inevitably realizing on our planet. So the further development of biology is impossible without the knowledge of the first stages of ontogenesis – embryology. That is why it appeared in the centre of attention not only of specialists-embryologists, but also of geneticists, biochemists, physiologists, cytologists, biophysicists and breeders, and is necessary basis for theoretical and experimental investigations dealing with the reproduction and its significance in the ontogenesis and evolution. Besides that, embryology disposes of a number of fundamental discoveries, which can be widely introduced into practice (cytoplasmic male sterility – CMS, phenomenon of apomixis, distant hybridization and some others).

Up to recent time breeders were unable to predict precisely the results of some or other crossing, its possibility to be put into practice and reproductive ability of the progeny. The processes occurring in flower – the fertilization, the development of endosperm and the embryo, the formation of reproductive organs were practically out of control. For scientists of the whole world more and more urgent becomes the question of new, untraditional approaches and methods permitting to reveal all potentialities of the plant organism, as well as to get new forms and sorts in shorter time.

In 1898 the Russian investigator, academician *Sergei Gavrilovich Nawashin* made an epochal discover – the double fertilization in flowering plants. It determined the further understanding of evolutionary process not only in angiosperms but also in all members of plant kingdom.

From the 30th of XX century the brilliant works on the fertilization process in various plant species, *Elena Nikolaevna Gerassimova-Navashina* begun that were later continued by her first learners – *cereals* (T.B.Batygina), *maize* (S.N. Korobova), *barley* (O.A. Dolgova), *orchids* (G.I. Savina), *onions* (I.A. Stozharova) made a breakthrough in this sphere. All they were realized on the high methodical level, almost all were cited in foreign editions and gained respect of the world scientific society.

Probably it will be appropriate to quote famous scientist N.V. Timofeev-Resovsky (1964), that *mysterious thing – "fertilized egg cell" occurs to be one of the most significant problems in biology*. According to his opinion, its studying is especially important to geneticists, because it is the source for different generations.

Almost a half of century passed after the brilliant lectures of N.V. Timofeev-Resovsky on the general problems of genetics but in present his idea is actual as well, especially in connection with the fact that the works on this topic unfortunately are practically absent, nevertheless in XXI century the Nobel prize was given for the investigations on human reproduction, in general related with early stages of ontogenesis.

It is quite evident that the possibility of regulation of separate stages of the development of embryonic structures, seedling as well as ontogenesis and life cycle in a whole is directly and closely dependent on the completeness of our knowledge of the normal morphological processes occurring in them.

The investigations on developmental biology already from the middle of XX century allowed to obtain the data on the morphogenesis of reproductive structures – ovule, anther and especially sexual embryo and embryoid (somatic embryo) etc. (V.G. Alexandrov, M.S. Yakovlev, T.B. Batygina, E.S. Teryokhin, G.Ya. Zhukova, V.E. Vasilyeva, O.P. Kamelina, G.E. Titova etc. see "*Comparative embryology of flowering plants*" in 5 volumes, 1981-1990 – The Government Prize in the sphere of science and technology was given to 8 collaborates of the laboratory, 1993).

Research on the cellular, tissue and organismic levels of all ontogenesis stages, beginning with organism arising turn to be a necessary theoretical basis for fundamental investigations in the theory of morphogenesis, morphoprocesses, genetics and breeding etc. (Batygina, 1950-2011; Batgyin, 1962-2000; Teryokhin, 1963-2001 etc.).

Any work connected with influence of *biotic* and *abiotic* factors even in early ontogenesis demands the knowledge of a whole picture of events taking place especially at early stages of ontogenesis including pollination and fertilization and determining its whole further course.

It is also impossible to use effectively the different methods of investigation such as *distant hybridization, breeding of hybrid embryos on synthetic nutrient medium, polyploidy, experimental mutagenesis, influence of ionizing radiation and conditions of breeding etc. without the deep understanding of embryonal processes*.

All above mentioned *has required* for investigators to reveal *the critical periods and stages* in ontogenesis (Batygina, 1974, 1987, 1999, 2010, 2011) that will facilitate the understanding of nuances in development and in some cases permit to control of definite ontogenesis stages.

I would like to attract a special attention to the fact that any changes in climate, anthropogenic factors disturb in plants from the very early developmental stages the reproductive structures first of all – egg cell, embryo, stamen, seed, in spite of their system of reliability (reserves), resulting in destroying and often disappearing of plant cover.

That's why in XX century the question on the necessity of elaborating of *special approaches and methods of preserving of biological diversity and resources* has been already discussed in the world biology.

In 1992 on the Conference of United Nations on the environment and development (UNCED) in Rio-de-Zanier (Brazil) the members of 178 countries including more than government chiefs discussed and signed five important documents. They are the Rio Declaration, determining the general principles of coordinating preserve of natural resources by different countries, Convention on Climate Change, Convention on Biodiversity, Statement on Forest Principles and Agenda 21, describing financial, technological and legal mechanisms for embodiment of actions on different aspects of environment preservation.

In this relation it should be noted that the preservation of wild plant species, restoring of soil and other seed banks, the possibility of obtaining in sort time of different forms and sorts etc. *require the knowledge of the whole picture of events, occurring from the moment of plant organism arising and determining the further course of individual development up to fruit and seed formation* (Fig. 1).

The golden age of biology, concerning its different trends – physiology, biochemistry, cytology, cytogenetics, genetics, breeding etc. reflected in the integration of all these directions.

Modes and types of plant reproduction and propagation

<table>
<tr><td colspan="3">With alternation of generations</td><td colspan="11">Without alternation of generations</td></tr>
<tr><td colspan="3">Sexual</td><td colspan="11">Asexual</td></tr>
<tr><td colspan="5">SEMINAL</td><td colspan="9">VEGETATIVE</td></tr>
<tr><td colspan="2">gamospermy</td><td colspan="3">agamospermy</td><td colspan="9">aspermy</td></tr>
<tr><td colspan="2">amphymixis</td><td>apomixis (gameto-phytic)</td><td colspan="2"></td><td colspan="9">amixis</td></tr>
<tr><td colspan="3">EMBRYOGENY</td><td colspan="5">EMBRYOIDOGENY</td><td colspan="6">GEMMORHIZOGENY</td></tr>
<tr><td></td><td></td><td></td><td colspan="2">floral</td><td colspan="3">vegetative</td><td>floral</td><td colspan="3">vegetative</td><td></td><td></td></tr>
<tr><td></td><td></td><td></td><td>embryonic (cleavage)</td><td>ovular (nucellar, integumentary)</td><td>foliar</td><td>cauli-genous</td><td>ризо-генная</td><td></td><td>foliar</td><td>cauli-genous</td><td>rhizo-genous</td><td>sarmen-tation</td><td>particu-lation</td></tr>
<tr><td>Capsella, Pisum, Paeonia</td><td>Avicennia, Rhizophora</td><td>Antennaria, Hieracium, Taraxacum</td><td>Eritronium, Orchis, Paeonia</td><td>Allium, Citrus</td><td colspan="3">Bryophyllum, Crassula, Ranunculus</td><td>Allium, Festuca, Poa</td><td>Hammar-bya</td><td>Lilium</td><td></td><td>Ajuga, Stachys, Paris</td><td>Aconitum, Carum, Plantago</td></tr>
<tr><td></td><td>Viviparity</td><td></td><td colspan="2"></td><td colspan="3"></td><td colspan="4">Viviparity</td><td colspan="2"></td></tr>
<tr><td colspan="3">EMBRYOGENIC</td><td colspan="5">EMBRYOIDOGENIC</td><td colspan="6">GEMMORHIZOGENIC</td></tr>
<tr><td colspan="14">types of reproduction and propagation</td></tr>
</table>

Figure 1. (from Batygina, 2000 with alterations).

A new discovery had for XX century – the stem cells and the questions connected with their development (totipotency, stemness etc.). The problem of totipotency, stem cells and the degree of stemness is considered to be one of the central in biology. The elaboration of a new science on stem cells requires the comprehension and involvement of knowledge from different disciplines.

I would like to attract your attention to the source of the problem on stem cells as well as to the importance of somatic cells in morphogenesis, ontogenesis and evolution: *stem cells* (Maximov, 1909), "*dormant meristem*" in leaf (Naylor, 1932; Yarbrough, 1932; McVeigh, 1938; Batygina et al., 1995, 1996, 2006, 2010) and ovule (Batygina, Freiberg, 1979; Batygina, 1991a, b; Batygina, Vinogradova, 2007), «*quiescent centre*» in root apex (Clowes, 1954), «*méristème d'attente*» in shoot apex (Buvat, 1955).

I consider *the creation of a new science on stem cells* to be examined in the aspect of developmental biology of plants and evolutionary biology of development (Evo-Devo). It should be mentioned that the progress in theoretical elaborations will be closely connected with the development of integration of various sciences and especially with *the problem of reproduction*.

I want once again to emphasize that in creating of a science on stem cells and their investigations the question on the development of its theoretical bases continues to be *actual*, taking into consideration comparative-evolutionary cellular biology for members of different kingdoms. To our opinion, developing the theory, we have to begin with elaboration of system approach (Batygina, 1983; Batygina, Vasilyeva, 1983), making possible the investigation of the *cardinal problem of the biology – the problem of wholeness*. This allows revealing correlations and studying correlative dependences of various structures at different levels of hierarchy – the above-organism, individual, organ, cell and molecular ones.

As it was truly regarded by the great Russian thinker-evolutionist I.I. Shmalhausen (1968), when studying all these processes there should be paid more attention to correlative interdependences of separate morphoprocesses and structures which, from the point of view of general biology, are considered one of the key moments in morphogenesis of plants – proliferation, differentiation, specialization, dedifferentiation, etc.

I suppose, that upon any comparative evolutionary researches of the developmental biology at various hierarchy levels, the location and properties of each structure should be performed taking into account *the centre of origin of species,* and also *its ethology, ecology and genetics* both for plants and animals (Batygina, 2010) (Fig. 5). It is important to take these peculiarities into consideration at the researches in the sphere of medicine – investigation of oncology etc. They determine reproduction strategy, plasticity, reserves and the system of reliability of the organism. The research of stem cells in culture *in vitro* should be based on *the regularities and laws of developmental biology revealed in natural conditions*. Comparative study of the morphoprocesses in natural conditions and in culture *in vitro* is one of the main positions in our experimental work.

The *primates* of deep investigation of any phenomenon *in nature* I always, from the studentship considered to be the main principle of her creative scientific work, because only in this case the phenomenon can be understood in the whole its complexity and diversity. As the great sculptor Auguste Rodin said, *Let the nature be your only goddess. Treat her with boundless trust. ... Be faithful to her without being afraid to lose your ambitions*.

In the present work the system of untraditional notions on ontogenesis, reproduction and plant evolution, based on the original results of multiyear investigations and analysis of literature in the area of developmental biology with using of the system integrate approach was suggested.

The creation of any science is impossible without applying of methodology. As outstanding theoretic of science and zoologist V.N. Beklemishev said (1964, 1994), *methodology is the part of logics*, one of the sciences studying *the human brain, the spirit* in its highest manifesting and the *technology of intellection*.

Besides that, the series of proposals, axioms and postulates lay in the base of all conclusions of any scientific branches, as V.N. Beklemishev considered. For the sake of clarity it is necessary to find out what their accuracy, what their and what their admissibility and what their evidence and usefulness! The special attention the author paid to the accuracy of notions.

The creation of the unified terminology for plants, animals and human taking into account their species specificity, is the serious problem, but each notion must be clear and correct (Van der Pijl, 1969; Batygina, 1987a-c; Beklemishev, 1994). This statement can be illustrated by the fact of misuse of such term as *«somatic embryogenesis»*, which in fact means formation of the embryo of uncertain origin – *somatic or sexual!* First of all, the term *«embryogenesis»* has been always used to designate formation of *the sexual embryo* as result of fertilization (*syngamy*) – heterophase reproduction. *The notion «somatic»* refers to embryo's origin – out of *somatic cell of various origin, i.e. to somatic embryo (embryoid) – homophase reproduction.*

We suppose, that it is necessary to use more correct term *"embryoid"*, which is the structural unit of a *new phenomenon* of vegetative propagation –

the phenomenon of *embryoidogeny*, and the process of embryoid formation should be referred to as embryoidogenesis (Fig. 2, 3). This phenomenon has a big significance in ontogenesis and evolution of plants (for details see: Batygina, 1977–2010). Probably, the term *"somatic embryogenesis"* has been long used in literature because of *misunderstanding of the essence of the notion.* It is an example of the fact, that at present time many of us face some conservativeness of thought and complexity of its reorganization. This issue had arisen once again when the new trends in the science appeared, the developmental biology and evolutionary developmental biology (Evo-Devo) side by side with the science embryology.

As Timofeev-Resovsky justly mentioned (1980), *the creation of a new direction in science requires the definite preparation of mind.*

Genetic and molecular analyses that have been undertaken during the last 50 years were mainly directed to revealing of the mechanisms of initiation of various structures, cell proliferation induction, systems of cellular signalization, etc. The important role of phytohormones in morphogenesis was confirmed by these studies, especially that of auxin, cytokinin and their ratios. A lot of information available today on plant stem cell regulation and on what transcription factors control cell proliferation and maintenance of stemness, as well as how this ultimately leads to differentiation in morphogenesis (Bosch, 2008; Lohmann, 2008; Batygina, 2009, 2010, 2011; Rudskiy et al., 2011).

Probably, the great Russian biologist N.K. Koltsov had a reason for believing that «... combination of ... sciences – *genetics and embryology ...*, as well as, *cytology and biochemistry,* will create a united science, which will allow solving general biological problems» (Koltsov, 1935, p.772).

Later he announced appearing of a new science – *developmental biology*, which was expected to study reasons and analysis of individual development. In 1967, in Moscow B.L. Astaurov had established the Institute of Biology of Development named after outstanding developmental biologist N.K. Koltsov, which studies various aspects of the life cycle, ontogenesis and evolution of individual representatives of plants, animals and of human beings.

Molecular genetics is one of the components of genetic-embryological researches. In the late XX the well-known geneticist M.E. Lobashov claimed *«... in order not to be in isolation from embryology genetics should seek for the efficient equipment for its researches which lies in discoveries of new genetic methods increasing the resolving capacity of the genetics analysis. This requires «intellectual equipment» – synthesis of biological knowledge and creativity»* (Lobashov, 1969).

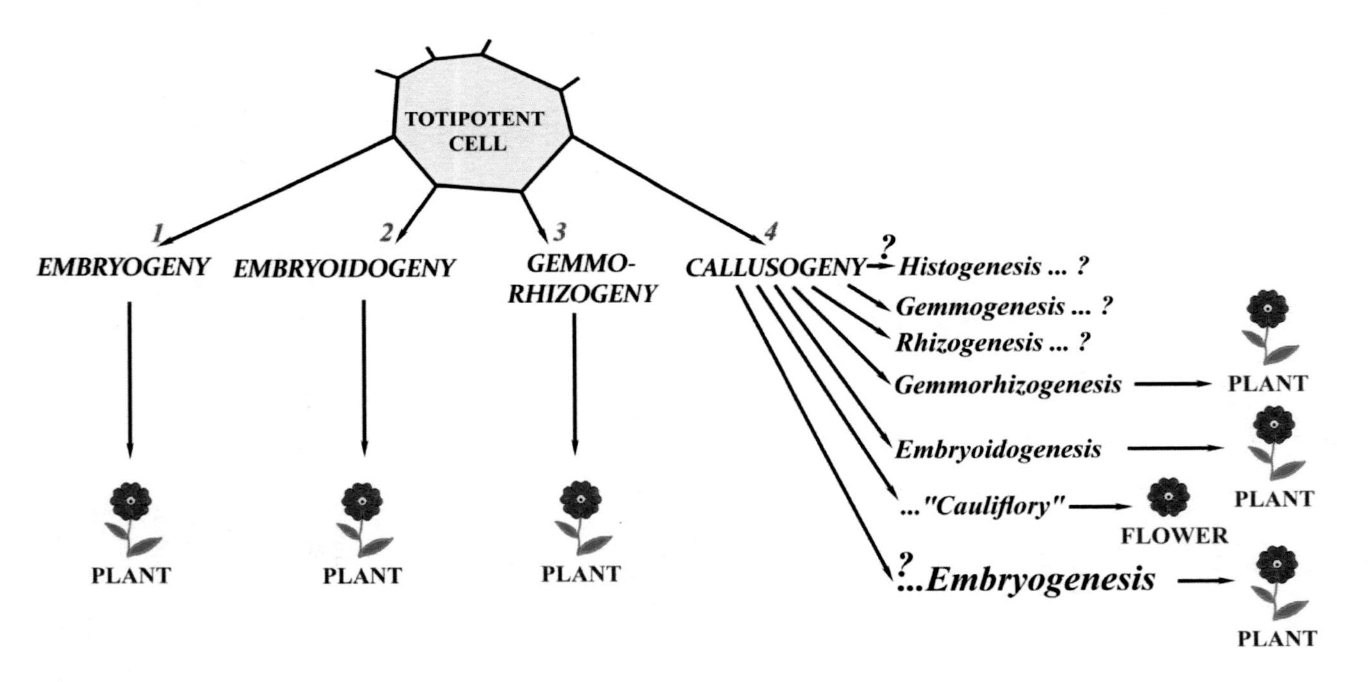

Figure 2. Phenomenon of embryogeny – 1 – sexual reproduction; 2 – embryoidogeny and 3 – gemmorhizogeny – asexual reproduction; 4 – callusogeny, different morphogenesis pathways are possible.

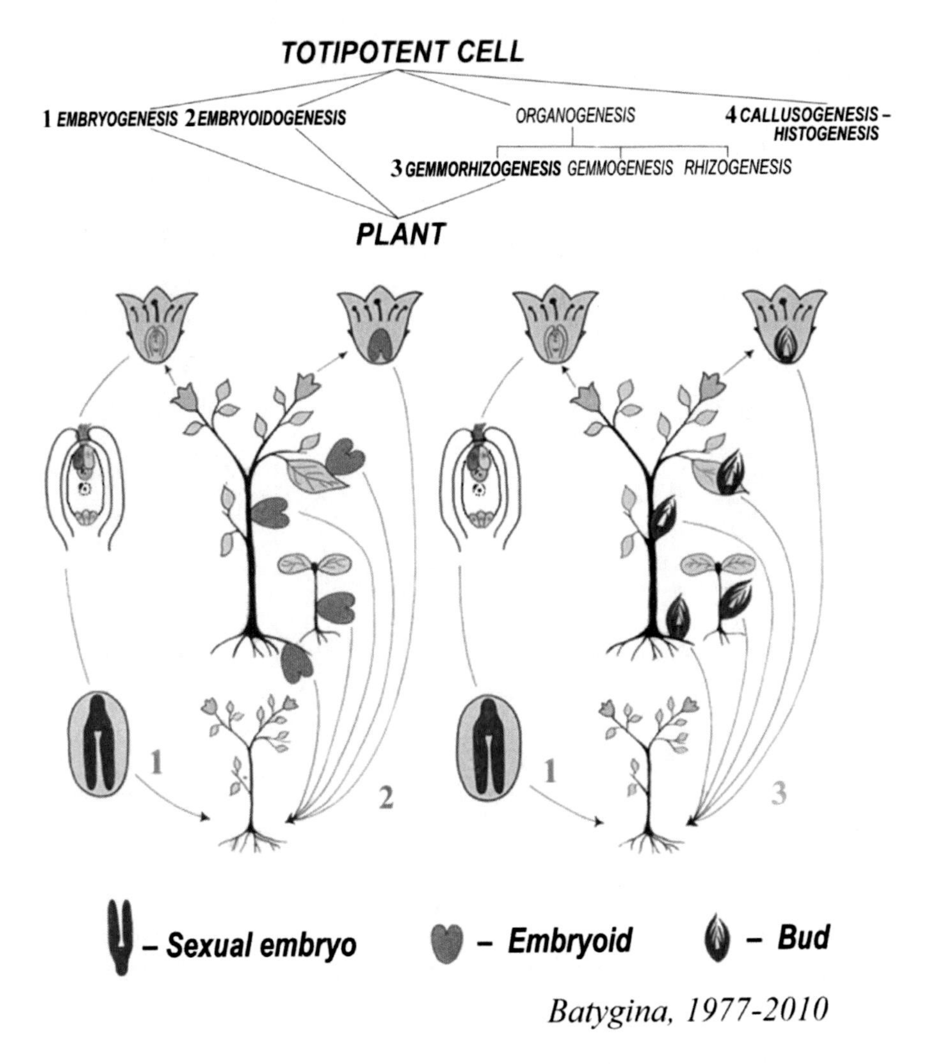

Figure 3. 1 – embryogenesis, sexual mode of reproduction; 2 – embryoidogenesis, 3 – gemmorhizogenesis and 4 – callusogenesis – histogenesis, asexual modes of reproduction.

Of course, in XXI century there are a lot of data, suggesting, that the *«intellectual equipment»* allowing us to join more and more with genetics is already found to a large extent.

During many decades the point of view existed that there is *a gap between sexual and somatic cells* and consequently between *asexual, vegetative propagation* (without meiosis and fertilization) where somatic cells in general occur to be the structural unit, *and sexual* one (with meiosis and fertilization).

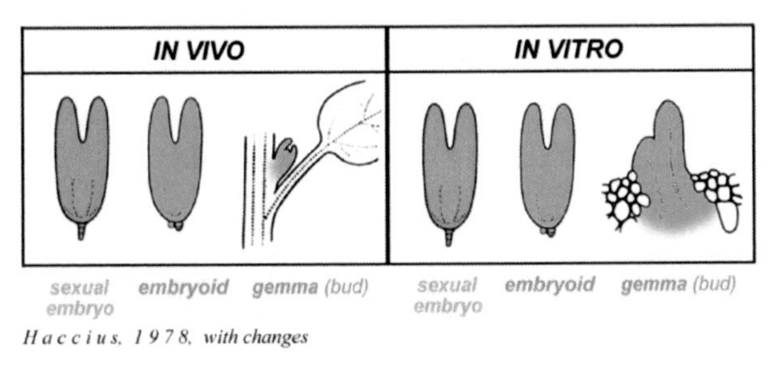

Figure 4. Various morphological structural units of sexual and asexual reproduction.

In this connection I would like to attract the attention to such discovers of XX century as growing plants out of somatic cells, haploidy, parasexual hybridization (Vöchting, 1906; Haberlandt, 1902; Butenko, 1964; Reinert, 1963 and others), which has stimulated works in the field of morphogenesis of various structures.

This has encouraged revealing of a new phenomenon, embryoidogeny – formation of embryoid – somatic embryo, a new type of vegetative propagation (Fig. 4). Embryoidogeny is found in representatives of various taxa in different ecological zones. Origin and location of embryoids on mother's plant determines the two forms of embryoidogeny: reproductive and vegetative (Batygina, 1977, 1978, 1984, 1987, 1993, 1994, 2002, 2006, 2007, 2009).

The unit of propagation in the embryoidogeny is embryoid - the primordium of a new organism, forming in the seed or at the vegetative organ, but not the part of the organism (or separate organs, as it is with graftage). Embryoidogeny were chosen by natural selection and appeared to be a central point in saving of the biological variety of plants. Geneticists and plant breeders make use of high potential ability of cells of nucellus of some plant species to produce nucellar embryoids in natural conditions, being the main criterion of genotypes selection for cross-breeding, which provides the variety in further generations.

Figure 5. Descriptive reproductive biology.

Besides that, *universality of pathways of morphogenesis* (embryogenesis, embryoidogenesis, gemmorhizogenesis) upon *individual formation* in *natural conditions* and *in vitro culture* was discovered (Fig. 2, 3) (Batygina, 1977-2010).

I believe, in all these discoveries, *the somatic cell* to be *the corner stone* both for morphogenesis and ontogenesis, reproduction and evolution. It is probable that we are able to discuss *the somatic evolution of cells.*

These discoveries have also favored further elaboration of the theory of reproduction and of regularities of morphoprocesses and their correlation dependences in the field of morphogenesis of reproductive and vegetative structures in plants. Especial attention was paid to *working out of stem cell theory* (the main properties of stem cells revealed, there has been developed the program of the study of cell lines and their participation in organogenesis with changing of stemness character and degree, etc.). Stem cells and their derivatives possess different potential – *the degree of stemness* depending on their function and niche (Batygina, 1979 – 2009a; Batygina et al., 2010; Rudskiy et al., 2011).

It should be noted that due to achievements of descriptive, comparative, experimental embryology and reproductive biology in XX-XXIth century a number of *discoveries* of general biological significance had been done. For the first time in the aspect of studying the cell cycle in various representatives of flower plants, there are discovered *the types of fertilization* (Fig. 6) (Gerassimova-Navashina, 1947-1978). The phenomenon of asynchronous transformation of the chromatin in male and female gametes upon the syngamy and triple fusion in cereals had been revealed (Batygina, 1961a, b), and subsequently — *M-type of caryogamy* (Fig. 7) (Batygina, Vasilyeva, 2002), movement of cells and nuclei in the microspore and pollen grain (Batygina, 1974).

There were discovered *new types of sexual embryogenesis* – *Graminad* (Batygina, 1968, 1969), *Paeoniad* (Yakovlev, Ioffe, 1957, 1983) and *sexual-somatic* (Batygina, 1977–2011; Brykhin, Batygina, 1994), *tubifloral type of development of endosperm* (Kamelina, 1997). The new genetical features of embryo – the *autonomity* (Batygina, 1987; Batygina, Vasilyeva, 2002; Vasilyeva, Batygina, 2006) and chlorophillness (Yakovlev, Zhukova, 1973; Zhukova, 1980) had been revealed. The mechanisms of genetic heterogeneity of embryos (5 types) and seeds (4 types) had been established (Batygina, 1993a), for the first time mechanisms of multiplicity of ontogenesis had been revealed in plants (Batygina, 2009).

At the end of the XXth century the approaches to studying new trends in embryology of flowering plants were developed – *the phenomenon of metamorphosis* in ontogenesis of high-specialized *parasitic plants* (Teryokhin, 1968, 1977, 1995), *population embryology* (1970), *ethology* (1972), developments of *the ontogenesis evolution theory* (1985-2001), *vegetative propagation of parasitic plants* –Teryokhin, Schuchardt, Wegmann, 1996.

The progress of the embryology in the 20th and 21st century has been caused by the use of system and complex approaches (Batygina, 1983; Batygina, Vasilyeva, 1983) and methods: temporal fixations, microscopy – the confocal one, Nomarsky optics, autoradiography, microsurgery, modeling, etc., resulting in revealing of specific peculiarities of meiosis with the help of endoenzyme markers, dimorphism of the sperms, in executing of fertilization *in vitro*, and in obtaining of unique data on cytoskeleton of reproductive and vegetative structures, etc.

MITOTIC HIPOTHESIS

Premitotic type | Intermediate type | Postmitotic type

Sea-urchin, et al. Fertilization in mature egg

Compositae, *Scilla*, Gramineae, *Fucus*, et al.

Lilium, *Fritillaria*, *Pinus* et al

Ascoris et al. Fertilization of immature egg

Elene Gerassimova-Navashina, 1947, 1982

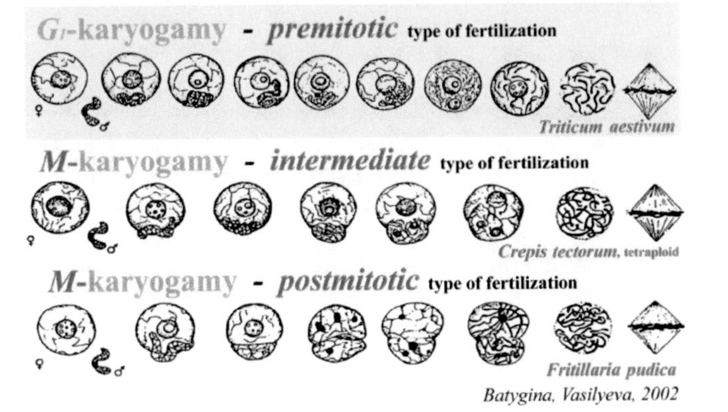

Figure 6. Mitotic Hipothesis.

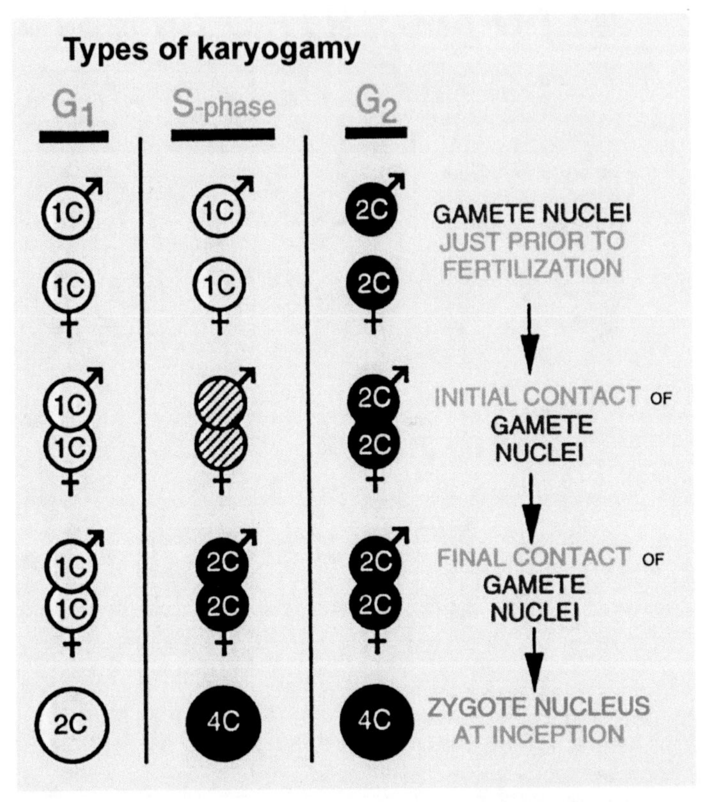

Carmichael, Friedman, 1995

FERTILIZATION type	PREMITOTIC			INTERMEDIATE	POSTMITOTIC
Type of KARYOGAMY	G_1-	S-	G_2-	M-	M-
Point of S-PERIOD beginning in sexual nuclei	after karyogamy	during karyogamy - end of nuclei fusion	before the beginning of syngamy	before the end of karyogamy, before the mitosis	
STAGE OF ZYGOTE DEVELOPMENT corresponding the S-period of cell cycle	MATURE ZYGOTE	IMMATURE ZYGOTE	GAMETES	IMMATURE ZYGOTE	

Batygina, Vassilyeva, 2002

Figure 7. Relationships of fertilization type, karyogamy type, cell cycle and stage of zygote development in plants; white circles – nuclei in G1-phase, hatched ones – nuclei in S-phase, black ones – nuclei in G2-phase of cell cycle.

Due to combined efforts of embryologists, geneticists and molecular biologists in the last decades of the XX and in the beginning of XXI century there have been identified specific genes, controlling flower development (ABC model) meiosis, development of egg cell and early stages of ontogenesis (Ezhova).

For the study of the problem of morphogenesis along with classical descriptive methods – cyto-embryological, cyto-histochemical, cytogenetic, electron-microscopic, etc., the experimental method including the culture of cells, tissue, organs and embryo becomes significant. The method is based on using of such unique property of cell as totipotency. When growing plant cells, tissues and organs there is set a task to restore in controlled conditions *natural course of their development to reveal morphogenetic mechanisms of fine regulation of life activity* at different levels of hierarchy.

This experimental method has allowed use various types of plant cells, tissues and organs as a model for *revealing potencies of plant cells.* This allows researchers to approach the controlling of the development of an organism.

Today developing the stem cells theory is within the science of developmental biology and evolutionary biology, and should be considered from the point of view of the integration of various sciences (Figure 5). It should be noted that for many years there have been studied various aspects of *descriptive and comparative embryology* taking into consideration classical notions, in particular, *the laws of embryogeny: of origin, numbers, position, destination and economy* (Souèges, 1937; Johansen, 1950), the revealing of *formation of the epyphisis and hypophisis* in the embryo of most of flowering plants at the early stages of development. This has become a considerable contribution to the study of *morphogenesis* of plants (Hanstein, 1870; Souèges, 1934; Raghavan, 1990). The *essence* of these structures lies in *organization of apical meristems of shoot and root of a future plant* (Batygina et al., 2004; Batygina, Rudsky, 2006).

Despite variety in structure of a shoot and, root, it is characteristic of all higher plants to have the quiescent centre or domain of non-differentiated cells in the apical meristem. Also, the time of their laying down is taxon specific and normally it happens in the fifth or sixth cell generation and is realized through asymmetric divisions. As a result, the whole structural domain of non-differentiated cells, as the initial of all following processes of organogenesis, by definition can be formed by cells of independent lines. *Diversity of known types of embryogenesis also demonstrates independence of arising of the shoot and root system organs. For example, in Graminad-type of development of an embryo both root tissues and shoot tissues can appear in very different cell lines* (Barlow, 1997, 2009, 2010; Batygina, 2005c, 2009b; Batygina et al.,

2010; Barlow, Batygina, orig. data). However, it is the cells of quiescent centre which are considered “stem cells” in traditional narrow sense of the term.

The term *«stem cell»* in the botanical literature, as well as in the zoological and medical ones, is used usually only as a *functional notion* (Batygina, Rudsky, 2006). The known complex WUS/CLV for stem cells of the shoot apex of *Arabidopsis* is only to some extent similar to QHB of root’s stem cells. Apart form general spatial location stem cells are also *joined by processes proceeding in the domain.* Relatively *symmetric divisions* are connected with self-maintenance of stem cells pool, while *asymmetric ones* – with *transition* of daughter cell to a certain way of *differentiation and specialization.* Stem cells derivatives – initials of cell lines – create the *niche of stem cells.* The surroundings of the domain of differentiating cells differs principally by presence of sister cells.

The investigations of the phenomenon of “stem cells” are considered to be one of the perspective directions in developmental biology. The Laboratory of embryology and reproductive biology of the Komarov Botanical Institute RAS, St-Petersburg, Russia, is working out for a long time the general biological problem of the origin and role of stem cells (dormant meristem) in ontogenesis of plant organisms (Batygina, 1979-2009a; Batygina, Freiberg, 1979; Batygina et al., 2004, 2010; Rudsky et al., 2011; Batygina, Rudsky, 2006; Batygina, Vinogradova, 2007; Titova, 2010).

The main properties of stem cells were elaborated, they demonstrate the wholeness of form-creation and reproductive processes at all stages of life cycle in plants (see in detail Part I).

1) *toti- or pluripotency*, i.e., the capacity for formation of *not only different tissue and organ types but a new individual as well*, by various morphogenesis pathways (*embryo-, embryoido-, gemmorhizogenesis*);
2) *selfmaintenance, i.e., the creation of cell pool generally owing to symmetrical divisions and the system of intercellular interactions*;
3) *the capacity to proliferation and formation of cells-ancestors* of different tissue types (“niches”)—owing to *asymmetrical divisions* under the definite signals;
4) *pulsatory and multistage character of the formation in tissue or organ*
5) *capacity to switch over the developmental program* that is provided by various molecular-genetic mechanisms (Batygina et al., 2004;

Batygina, Rudskiy, 2006; Batygina, Vinogradova, 2007; Batygina, 2005a, b, 2007a-c, 2010, 2011).

System approach and analysis of data on "maturation" of male and female gametes (egg cell at different stages of development), double fertilization in model objects: syngamy, triple fusion, plasmogamy and karyogamy taking into consideration the cell cycle, zygote, embryogenesis of model objects in representatives of a lot of flowering plants taxa (Fig. 8-14) allowed us to express from the embryological point of view the nontraditional idea: the zygote is proximate stem cell (stem cell of 1st order) and gives a rise for all other stem cells (of the next orders) in plant ontogenesis, particularly those of shoot and root apices.

Multi-stage transformation of egg cell into zygote

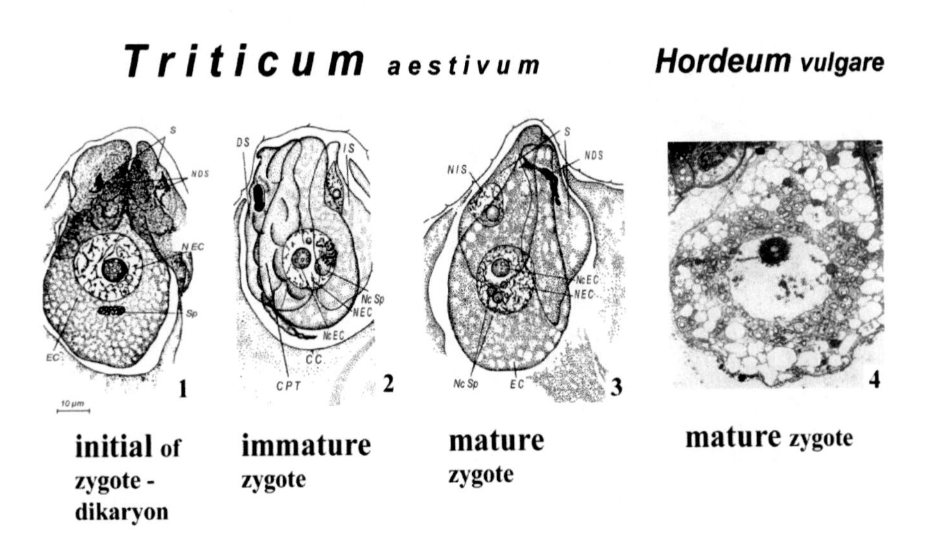

Figure 8. Syngamy. Micropylar part of embryo sac in 1 h – (1), 5 h – (2) and 7 h – (3) after pollination. EC – egg cell, S – synergid, N D S – nucleus of degenerating synergid, N EC – nucleus of egg cell, Sp – sperm, D S – degenerating synergid, I S – instant synergid, C P T – content of pollen tube, C C – central cell, Nc EC – nucleolus of egg cell, Nc Sp – nucleolus of sperm, N I S – nucleus of instant synergid; (1-3) from Batygina, 1974-2000; (4) from Norstog, 1972.

Every plant cell is able to be in different morphophysiological states, the characteristics of one of them mostly correspond to the properties of stem cell, the others correspond to actively *dividing meristematic cell.*

The transitions between these states are stochastic and in most cases histogenesis and organogenesis are connected with the loss of so called stemness, but any other transitions are also possible (Barlow, 1997; Batygina et al, 2004).

The study of descriptive, comparative, experimental developmental biology and other integrative trends (see Figure 5), and especially comparative embryology of representatives of various taxa of *flowering* plants enables to consider that *formation of stem cell of orders derived from zygote, are probably typical for all organs – flower, stem, leaf, root – and at all stages of the life cycle – sporophyte, gametophyte, their functioning, first of all, depending on their localization and function.*

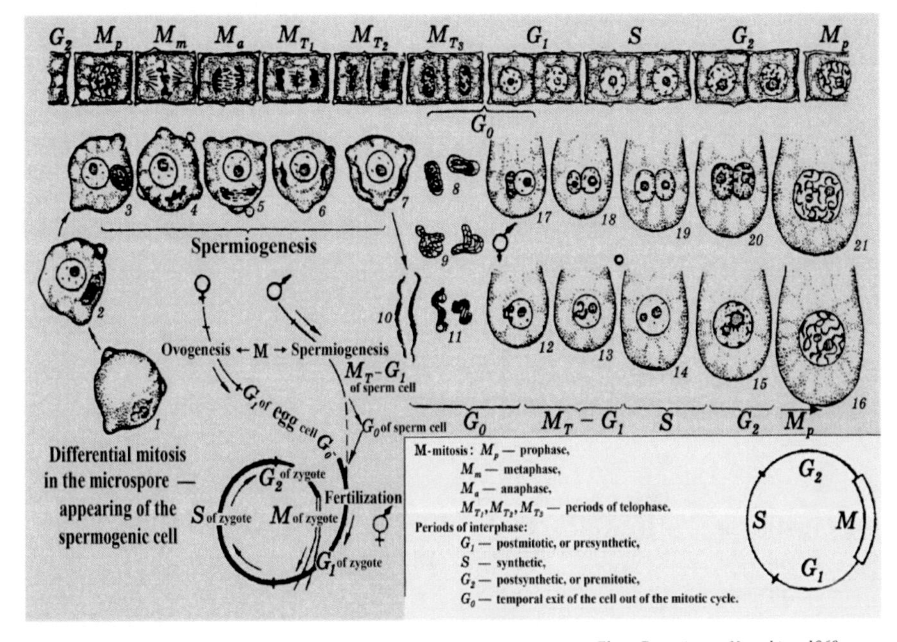

Figure 9. Changing of sexual cells during fertilization in correlation with periods and phases of mitotic cycle: (1-11) – spermiogenesis; (12-16) – premitotic type of fertilization; (17-21) – postmitotic type of fertilization; the upper row of figures – mitotic cycle of meristematic cell.

It is from the point of view of stem cells that *initial cells of somatic embryos (embryoids) of various origin in natural conditions and in culture in vitro* should be considered. In this connection, it is of a great importance to study *the new phenomenon of embryoidogeny – a special category of vegetative propagation* (Batygina, 1977-2010).

Progamic phase — flowering and pollination

Triticum sp.

Figure 10. (1, 2) flowers opening in middle part of ear, (3, 4) stamens (anther+filament) emerging and anthers opening, (5, 6) pollination, (7) germinating pollen grains (*pg*) on stigma (from Batygina, 1987a).

***Progamic** phase of fertilization in cereals*

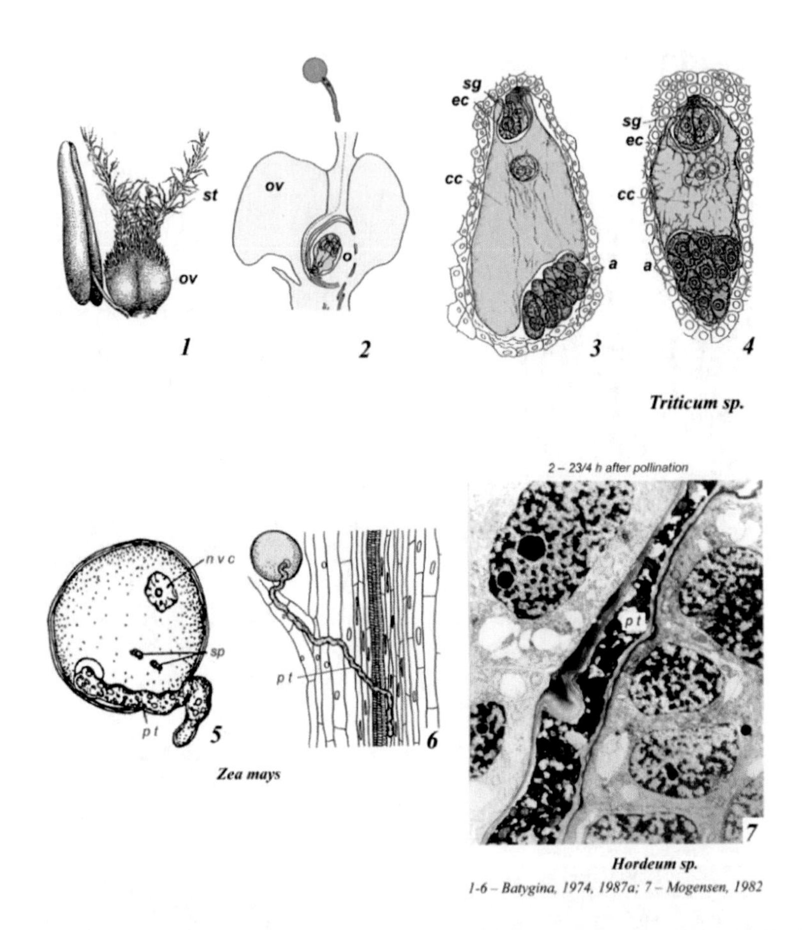

Figure 11. (1) general view of pistil and stamen; (2) pollen tube way (dotted line) in ovary; (3, 4) mature embryo sac before fertilization – side view (3) – and full face – (4); (5, 6) pollen tubes germination on stigma and style; (7) in micropylar area between ovary wall and inner integument, there can be seen degenerated cytoplasm of pollen tube.

a – antipodals, *cc* – central cell, *ec* – egg cell, *n v c* – nucleus of vegetative cell, *o* – ovule, *ov* – ovary, *pg* – pollen grains, *pt* – pollen tube, *sg* – synergids, *sp* – sperm cells, *st* – stigma. (2-6) – light microscopy, (7) – TEM.

For the first time it is grounded, that the zygote in plants (Fig. 8) – an initial cell of sexual embryo – is a unique differentiated totipotent stem cell, «progenitor» of stem cells of other lines. It contains information about architectonics and development of organization of an individual, connected with gradients and correlations between its parts, and consequently with heterogeneity of different axes of a future organism (Batygina et al., 2004).

Syngamy and triple fusion in Triticum sp.

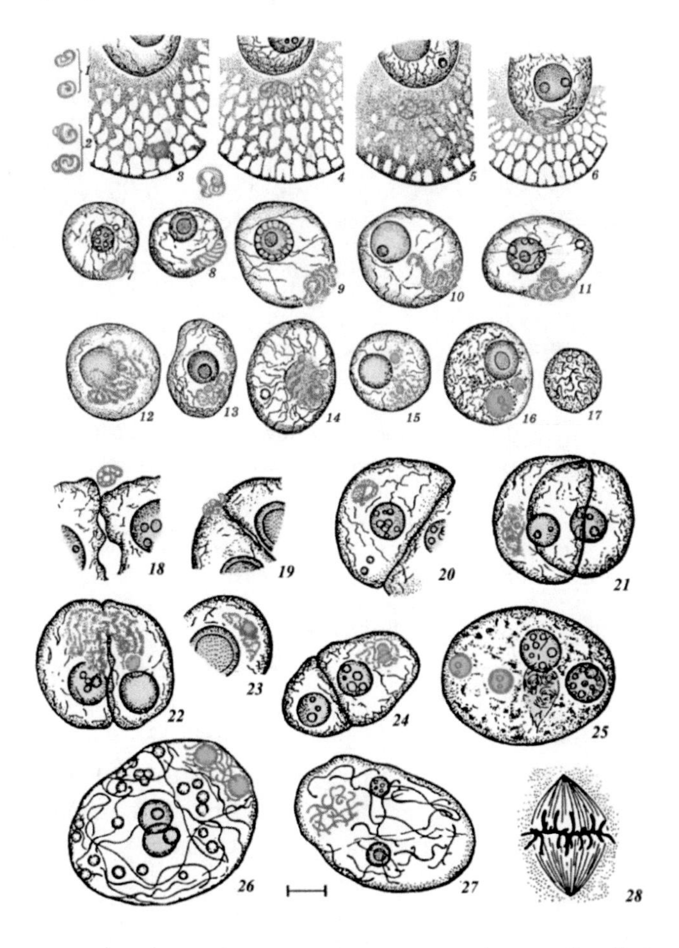

Figure 12. (1-17) sperm from entering embryo sac up to its complete fusion with egg cell nucleus: (1) sperms from synergids, (2) sperms from "slot", (3-6) sperms entered egg cell, (7-15) fusion of sperm and egg cell nuclei, (16) zygote nucleus in interphase, (17) prophase of zygote; (18-28) second sperm from entering embryo sac up to its complete fusion with polar nuclei: (18) sperm approaching polar nuclei, (19, 22) sperm bordering two polar nuclei; (20, 21, 23, 24) sperm fusing with polar nucleus; (25) a brief (5-10 min.) "interphase"; (26, 27) transient stages of unition of two polar nuclei, separate entering of polar nuclei and sperm into prophase; (28) metaphase – complete unition of sexual gametes.

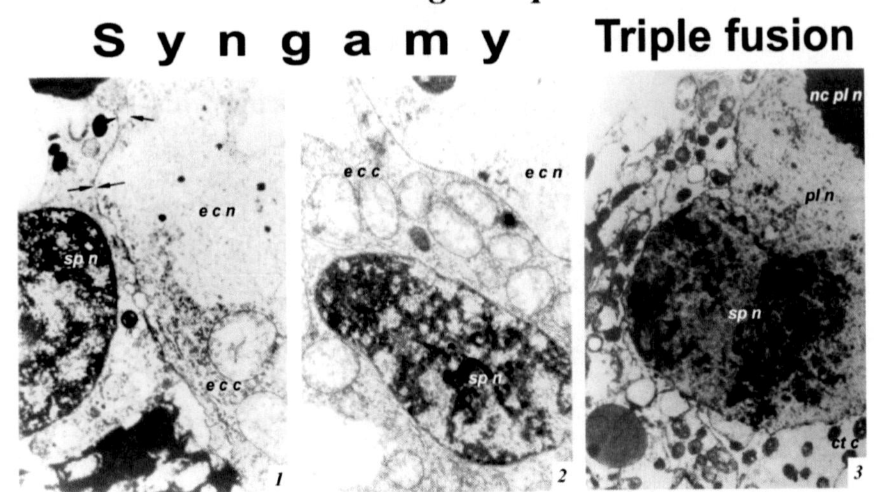

Figure 13. Syngamy and triple fusion in *Hordeum sp.:* (1) contact of egg cell and sperm cell plasmalemmas; (2) sperm cell nucleus in cytoplasm near egg cell nucleus; (3) – triple fusion – sperm's nucleus, fusing with the polar nucleus.

sp n – sperm's nucleus, *e c n* – egg cell nucleus, *pl n* – polar nucleus, *nc pl n* – nucleolus of the polar nucleus, *e c c* – egg cell cytoplasm, *ct c* – central cell.

I would like to cite here the words of N.V. Timofeev-Resovsky said almost half of the century ago: «Every species is presented through time by almost infinite alternation of growing and dieing generations. And here is the main problem, which we have to raise in regards of the genetics (I am, particularly, suppose it to be the central problem of the biology in general): it comes to the fact, that each generation is separated from another one by a single cell stage. Here is an individual, and here is another one of the next generation, and they are connected with the stage of a single cell, the fertilized egg cell».

For my opinion, today these words sound visionary, because the problem of fertilized egg cell – zygote, the initial and proximate stem cell of the new organism, in all representatives of different kingdoms becomes the central problem of evolutionary developmental biology (Evo-Devo).

Ii is known, that the totipotent stem cell, the zygote, which is a result of sexual process – fusion of two gametes, develops into single multicellular organism.

Double fertilization in Torenia fournieri

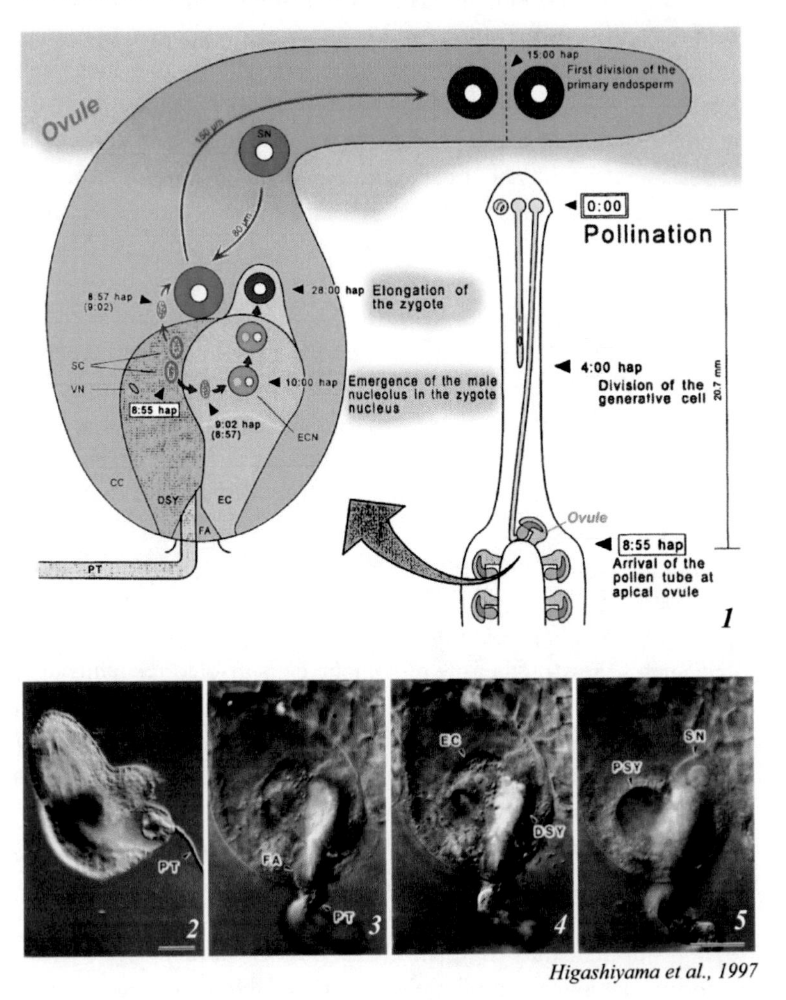

Figure 14. (1) Time course in pistil – right, and in naked embryo sac of apical ovule – left; (2) general view of ovule with a naked embryo sac penetrated by a pollen tube (10 h after pollination); (3-5) extramicropylar area in different focal planes; (2-5)-DIC.

hap – hours after pollination, CC – central cell, DSY – degenerated synergid, EC – egg cell, ECN – egg cell nucleus, FA – filiform apparatus, PSY – persistent synergid, PT – pollen tube, SC – sperm cell, SN – secondary nucleus, VN – vegetative nucleus.

The example of what Timofeev-Resovsky asserted is the monozygotic twins (or triplets, etc.), produced in some cases in plants and animals upon the gametes fusion or, possibly, at later stages of organism development. Strange as it may seem, but these individuals form as result of asexual formation of the organism (vegetative propagation) and change of morphogenic program from sexual to asexual process. In present the clear data on the mechanisms of formation of a single or several organisms from zygote are absent.

The phenomenon of the formation of monozygotic twins proves the fact that it is possible to speak about the homology of sexual and somatic cells.

Not only zygote but also somatic cell is capable to be the initial and the first stem cell of a plant.

The elaboration of the reproduction theory has revealed the essential role of somatic cells in ontogenesis and evolution. It is rather obvious for embryologists that at sexual and asexual (from the somatic cells) reproduction the organisms possess the information on different aspects of organization, initiated on different structural base. The data on the general architectonic exists mainly on the level of zygote or somatic cell.

New individual is born independently from the origin: from zygote or somatic cell.

The phenomenon of embryoidogeny demonstrates the significance of the role of somatic cells in organization of embryo body, in reproduction, particularly in changing of morphogenetic program in ontogenesis and evolution. Such a view on the role of somatic cells also corresponds to the data of the following five discovers made in the middle of XX century: stem cells, growing plants from somatic cell, haploidy, parasexual hybridization and embryoidogeny. Besides that, such structures as dormant meristem, and dormant center were revealed.

Embryologists once again confirmed that no gap exists between sexual and somatic cells, and permitted to discuss the question on the possibility of homologization of some totipotent somatic cells and zygote.

Progress in the study of stem cells biology and the phenomenon of embryoidogeny is connected with elaboration of the theory on developmental biology and the theory of reproduction in ontogenesis and evolution. The integration of all trends in research of cells and embryoidogeny from the point of view of wholeness at various levels of biohierarchy will enable developing a number of hypotheses and lawss, as well as the further working out of basics of the stem cells theory, and formulating new laws of development taking into account the existing data of classical works and basing on general regularities, existing laws of biology, morphogenetic correlations considering ethology and ecology.

The fundamental investigations in the theory of stem cells is one of the bricks in the construction of the theory of developmental biology and the theory of reproduction.

The knowledge of the theory on reproduction allows predicting the genotype of progeny and opens new perspectives in controlling of certain ontogenesis stages. Embryological information is rather important also in creation of original selection programs and industrial technologies with untraditional modes of breeding of new plant forms and sorts.

To our opinion, in present the following question turns to be n of the most important: how far and deep we could move in the investigation of excursively interesting and significant phenomena determined the progress of understanding of planet living beings, e.g. conservational biology.

Embryology of plants was born as a discipline of botany, during XIX-XXI centuries has undergone great changes and passed a long way from description embryology to developmental biology – the science on the regularities of genesis, development and further ontogenesis stages.

Developmental biology – a new area of biology, reflecting all phenomena connected with the development of form and function and, naturally, ontogenesis evolution and characterized by integrativness.

New direction of developmental biology – evolutionary developmental biology studying the evolution of ontogenesis mechanisms in English literature obtained the shorten name "Evo-Devo" and it has absolute relevance (Dondua, 2005).

I will allow myself to offer my colleagues my considerations and some developed provisions – preconditions of the developmental biology and Evo-Devo, and it is in this aspect, that the development of the stem cell theory will be viewed (Batygina, 1977-2011).

Preconditions:

- Moving of embryology onto a brand new level: from *the descriptive, comparative embryology* to the *developmental biology* – morphogenesis, structure, function, and naturally, the evolution of ontogenesis.
- Learning the individual development and *potencies of an organism*, knowing of the essence and reasons of *morphological processes* and *morphogenesis* (Beklemishev, 1994), revealing genetic and molecular mechanisms of initiating of various structures, induction of

proliferation, cell signaling system and other is a new trend in biology – *evolutionary developmental biology, «Evo-Devo»*

- *Arrangement of knowledges, integration* of traditional and non-traditional *scientific trends* and developing new ones enables discoveries of general biological importance (Batygina, 1978-2011).
- *Integration and formation of " "interdisciplinary trends"* bridging traditional specialisations and thus enabling the creation of *the universal science* to serve as a kind of frame *bringing together* separated sciences.*" (Takhtajan, 1998, p.11)*
- Creation of a new scientific trend requires a certain preparation of the mind (Timofeev-Resovsky, 1980), methodology studying the human mind and spirit in its highest manifestations, and the technology of reasoning (Beklemishev, 1994).
- *"To understand each other"* a special attention should be paid at preciseness and clarity of the notions – hypotheses, postulates, axioms underlying any conclusions of every trend of each branch of science, their evidence, legitimacy and productivity (Beklemishev, 1994; van der Pijl, 1969; Batygina, 1999).
- Attracting famous scientists, as well as researches who are able, from their early ages, to develop nonconventional thinking identifying the priority of trends, and accompanied with *intuition*, broad *expertise* and *fundamental knowledge*, and who are not afraid of "conservatives" will allow to enhance the integration with other scientific branches and trends.
- Development of the reproduction theory (Batygina, 1974, 1977, 1987, 2010, 2011; Batygina et al., 1978; Batygina, Vasilyeva, 2002).

The Basic Principles of Reproduction

The main statements of the reproduction theory have been developed and formulated. They were based *on the principles of formation of reproduction systems, enabling plasticity and tolerance (adaptation) and determining the reproduction strategy of the species in ontogenesis, life cycle and evolution*:

- the principle of the alternation of haploid and diploid generations, universality of morphogenesis pathways, status and interaction of types, modes and forms of seed and vegetative propagation;

- the principle of structural organization, space-temporal coordination, critical periods and polyvariety of development;
- the principle of wholeness, integrity and sustainable development from the point of view of reliability system, "doubling mechanisms", *reserves* and *failures*, which are characterized by various degree of oligomerization at different levels of the hierarchy, and do not allow the complete failure in the system of plant reproduction;
- the principle of self-regulating dynamic integrated system of an organism, determined by the property of *totipotency* and *stemness* showing itself at different time and at different rate, and enabling *continuity of morphogenesis*, a range of modes – *sexual, asexual, apomixis*, and forms of reproduction – *viviparity* and other, as well as *pathways of morphogenesis* accompanied by polyvariety of matrix processes and heterogeneity of cells;
- the principle of universality of phenomena: totipotency, stemness, multiplicity of ontogeneses, *metamorphose, embryoidogeny, viviparity, species specialization, possible repeatedness of their changes and switching of various morphogenetic programs of development*

sexual ↔ asexual,

gametophyte ↔ sporophyte,

♂ gametophyte ↔ ♀ gametophyte.

The arsenal of the data received in the XX-XXI centuries on biology of development including stem cells biology has allowed me to propose and discuss the three main perspective trends in study of stem cells:

- gametes – zygote – embryo – individual – population – coenosis;
- different types of somatic cells considering their localization and function in the organism; possibility of homologization of some somatic cells with the zygote;
- apical meristem of shoot apex and root apex in which organization somatic cells take part.

Chapter II

The Modes of Organism Formation, Propagation and Renewal

General Idea

During global climate change, anthropogenic factors firstly damage the reproductive structures in plants (stamen, embryo, and seed), resulting in impairment and often disappearance of vegetative cover. In this connection, the preservation of the natural plant gene pool, restoration of soil seed bank and finding new forms and sorts in the shortest terms require knowledge of the whole perspective of events that take place from the origin of the plant organism and determining the course of individual development up to the formation of fruits and seeds.

The systems of plant reproduction are considered to be examined both in relation to the character of genetic information transferred to posterity and the modes of its realization in ontogenesis. Due to the factors of environmental influence, the adaptive changes occur not only in genetic information but also in the modes of its transferring.

All processes resulting in the increase of biological units are referred to as propagation. Propagation can take place at different levels—molecular, cellular, tissue, organ, organism and population. Cell division is the basis of propagation (cytogony). According to traditional classification, plants have *three types of propagation: sexual, asexual, and vegetative*. In spite of a large number of investigations devoted to different aspects of propagation phenomenon, the biological entity of definite types, modes and forms of reproduction and their interactions are considered to be not clear enough.

The question on the volume of notions of *"sexual process"* and *"sexual propagation"* is also debatable. All these terms require definition and, as L. van der Pijl noted, "...differentiation of terms is *not rather the play of words* but an absolutely *essential condition* in order to understand the nature of things" (1969, p. 15). We tried to express our *non-traditional view of this problem.*

On the example of the notion *"sexual process"* one could easily see how much the terminology, concerning propagation, is mixed up. The sexual process in plants in its typical form is understood as the fusion of two sexual cells – gametes and the formation of zygote (Figure 8, 12-14). Interpreting the notion "sexual process" in flowering plants most investigators ignore one of its key periods – meiosis. From our point of view *the sexual process includes meiosis and fusion of gametes (from different meioses), which results in arising of a zygote, i.e., a new individual.* Thus during sexual process the increasing number of individuals does not occur, since a single new organism forms, as a rule, from the zygote, developing as a result of the fusion of male and female gametes. *The enlarging of sexual progeny* is provided just by *the multiplicity* of generative structures (ovules, pollen grains, gametes, zygotes) and so by the multiplicity of sexual processes. That is why, when *only one sexual process* takes place, *reproduction* should be mentioned.

The term *"sexual process"* is often replaced in literature by the notion *"sexual propagation"* which, according to the notion mentioned above, is not correct, and these terms could not be considered synonyms (see Batygina, Vasylieva, 2002).

The famous Italian botanist, Battaglia (1963), while examining the phenomenon of reproduction from the general biological view, formulated the notion that organisms are able to arise *in the same phase—sporophyte from sporophyte (homophasic increasing in number, "multiplication")—or, in the alternative (antithetic) phase, sporophyte from gametophyte (heterophasic increasing in number, "reproduction").*

It should be mentioned that one speaks about asexual type of propagation in different members of plant kingdom "in the narrow sense"—*multiplication by spores* and "in the broad sense"—*vegetative propagation*. Multiplication by spores is typical for algae, Bryophyta, Licopodiophyta, Equisetophyta and ferns. Pisyaukova (1980) suggested using *for the flowering plants the term "vegetative multiplication" only*, considering the usage of the term *"asexual multiplication"* to be in general tautology. According to her opinion the spores of flowering plants *lost their main functions, connected with multiplication and dispersal.*

Thus, the analysis of literature and original data *allows decline the terms "sexual propagation" and "asexual propagation" related to the flowering plants, exclude them from the generally accepted classification of propagation types and use instead of them the notions "seed propagation" and "vegetative propagation".*

In the native and foreign literature ones often use the term "reproduction" (French *reproduction*, German *reproduktion*). It should be noted that only the detailed analysis of article context, where this term applied, permits understanding of the meaning utilized by the authors—reproduction, propagation or renewal.

All questions related to the problem of *propagation* appeared to be *pure rhetoric* at first sight. But unfortunately, as we satisfied, *interpretation of many notions does not reflect the essence of phenomena both from genetic and general biological positions.* The debatable character of the notions *"propagation"* and *"renewal"* is also considered to be the confirmation of this. It should be mentioned that the first applies to the individual and the second to the population. The term "propagation" is not applied to the systems of above organism level. At such distinguishing of the notions the "vegetative renewal" is understood as the renewal of population by vegetative propagation (Levina, 1981). Accordingly, the "seed renewal" is understood as the renewal of population by seed propagation.

Seed propagation, as well as vegetative renewal, depends on numerous different environmental factors: meteorological, edaphic, allelopathic and therefore occurs to be the *stochastic process.* However *sexual* and *asexual processes* of individual formation, *heterophasic and homophasic reproduction, seed and vegetative propagation* and *renewal* elapse conjugately, the certain correlations, which permit to remain *the homeostasis of species and population*, are revealed between them. But the question on *the ratio of different types, forms and modes of propagation in various taxa believe to be the most difficult and discussible.* The solution of this problem has a great importance for the study of plant adaptation, preserving of biological diversity and biological resources as well as for the elaboration of biotechnological approaches for the mass copying of value genotypes.

Seed Propagation

The flower is the most important reproductive organ of angiosperms; in its ovary, the processes of *micro- and megagametogenesis, pollination and fertilization* occur, resulting finally in *the formation of fruits and seeds containing a new sporophyte, the embryo*. The flowers are extremely diverse in structure and colour, their diameter varies from millimeters (for example, orchids) up to the metre and more (for example, *Rafflesia*).

The flower is the totality of sterile and fertile structures developed on the receptacle. The *sepals and petals* belong to the sterile structures; *stamens and carpels* belong to fertile ones (Figure 15). The borders between sterile and fertile structures are conventional in some plants (Figures 16-18).

In the classical concept, the members of different groups of plants (mosses, club-mosses, ferns, gymnosperms and angiosperms) undergo in their life cycle the alternation of generations: *sporophyte and gametophyte*. The term "alternation of generations" was introduced by Hofmeister (1851). Meiosis occupies the central place in reproductive cycle, at the alternation of *diploid and haploid phases*. From each cell with the diploid chromosome number the *haploid* sexual cells (gametes) are produced in the course of meiosis. Haploid stage of plant development is represented by *gametophyte* (beginning from the growing of micro- and megaspores up to zygote). An organism developed *from zygote up to the producing of spores is referred to as the sporophyte*. In the process of evolution towards angiosperms, the gradual reduction of gametophyte and domination of sporophyte occur. The sporophyte of angiosperms is represented by trees, bushes and herbs. The gametophytes are reduced and are formed as a result of several mitotic divisions. The development of an *embryo sac* (female gametophyte) is diverse, however in majority of plants it is usually reduced to three divisions (Figure 19-22), and the development of *pollen grain* (male gametophyte) to two.

V. A. Poddubnaya-Arnoldi (1976) considered the phenomenon of "alternation of generations" from a different point of view. She assumed that sporophyte and gametophyte in angiosperms are so much connected with each other, that *they should not be examined in isolation with each other since the plant is a integral organism.* One could speak only about *the developmental phases in the life cycle of plants*. However, as R. E. Levina (1981) emphasized, the introduction of the notion "alternation of generations" has played a positive role in science, as it availed to reveal the unity of origin in different plant groups. We share the opinion of V. A. Poddubnaya-Arnoldi and R. E. Levina. *Theoretical elaborations on the research of polyembryony phenomenon from the positions of the reliability system of the organism (plasticity and tolerance)* allow us express the idea that *the dividing on sporophyte and gametophyte in flowering plants is conventional* (Batygina, Vinogradova, 2007).

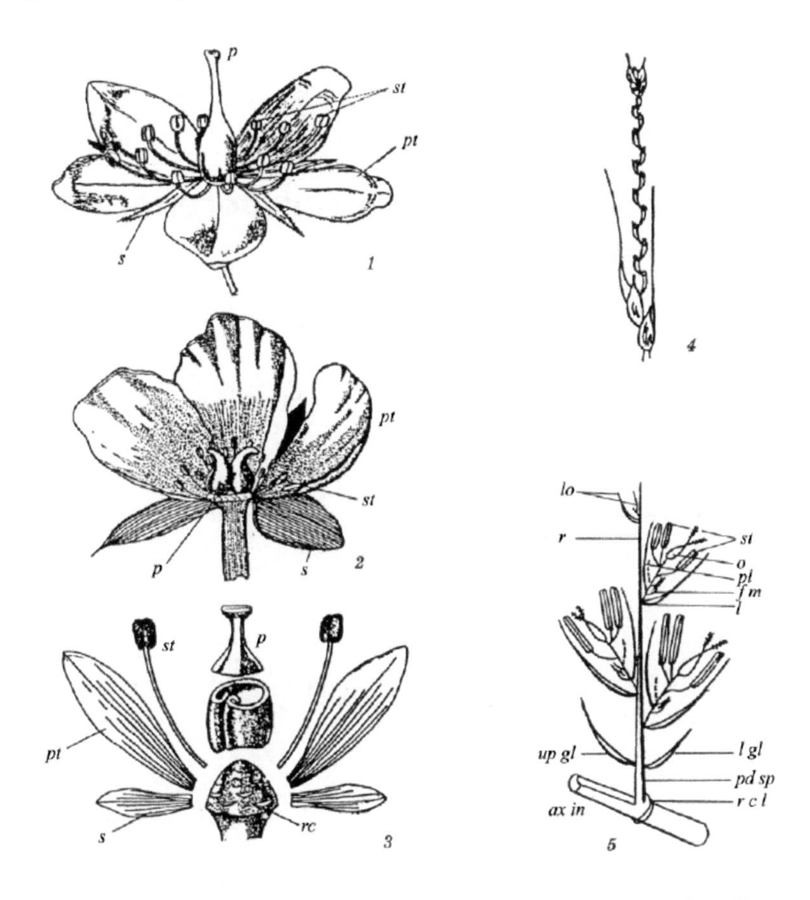

Figure 15. Structure of the flower and inflorescence: (1) general view of the flower; (2) the peony flower (in longitudinal section with a lot of stamens and two pistils; other have been removed); (3) separate parts of the flower; (4, 5) the structure of the ear (4) and triflorous spikelet (5) in cereals;

up gl – upper glume; *pl* – plea; *o* – ovary with two stigmas; *lo* – lodicule (flower glume at upper half-grown flower); *pt* – petals; *pd sp* – pedicel of the spikelet; *l gl* – lower glume; *l* –lemma; *r* – rachilla; *ax in* – common axis of inflorescence; *p* – pistil, *r c l* – rudiment of the covering leaf; *st* – stamen; *f m* – flowering membrane; *rc* – receptacle; *s* – sepal; (1-3) from Kursanov et al., 1940; (4, 5) from Batygina, 1987a.

While examining the question of seed propagation, we should once again mention that the peculiar features of reproductive life in plants are to a great extent conditioned by their immobility. The formation of *special structures* (*diaspores*) which make easier the species propagation and dispersal became one of the consequences of this. In the course of many millions years of plant world evolution the spore stay to be universal diaspore (algae and higher cryptogamous). Later, in seed plants, this function transferred to *the seeds* that helped them to occupy the dominant position on the land.

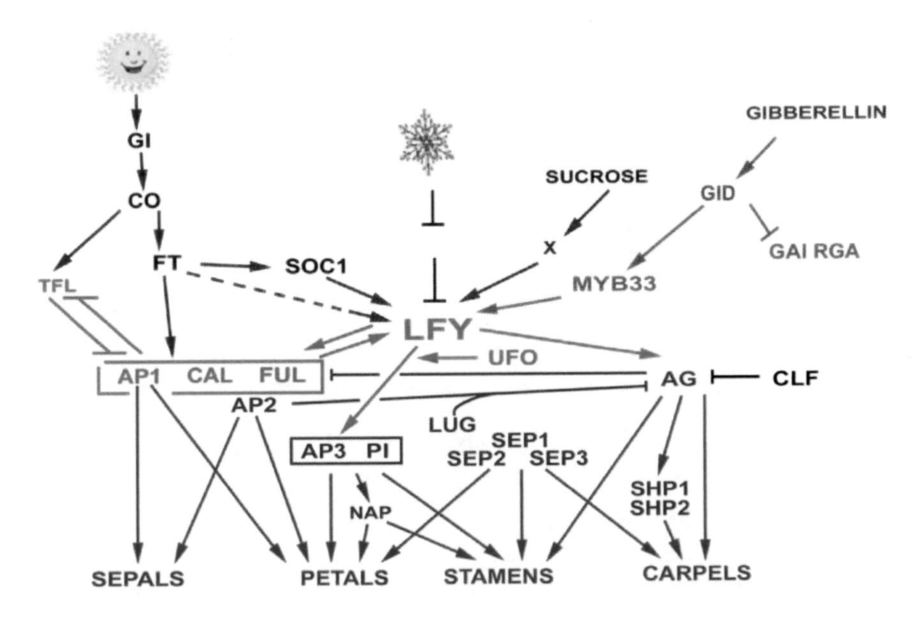

Figure 16. Integration of different inductive signals of blossoming by *LFY* gene. Green color shows the main genes controlling the development of the floral meristem and flower organs. Other colors show genes initiating blossoming in response to induction with light (photoperiod), cold, hormones and sugars (from Ezhova, 2010).

The seed, one of the units of propagation and dispersal typical for seminal plants, contains *the initial of new plant — the embryo —* and specialized storage tissue (endosperm, perisperm, etc.) enclosed into the seed coat (testa, spermoderm) (Figure 23).

Seed propagation in plants is provided by the account of formation of big number of seeds with embryo. One should note that in the same seed two processes of new individual producing can take place: *sexual* and *asexual*. Owing to this, the seed can contain *several embryos of different origin* (*polyembryony*) and give the seedlings of different genetics (*matroclinous posterity* (2n=2n) — from nucellus and integuments) or *the individuals* (*with new genetics n+n – from zygote*).

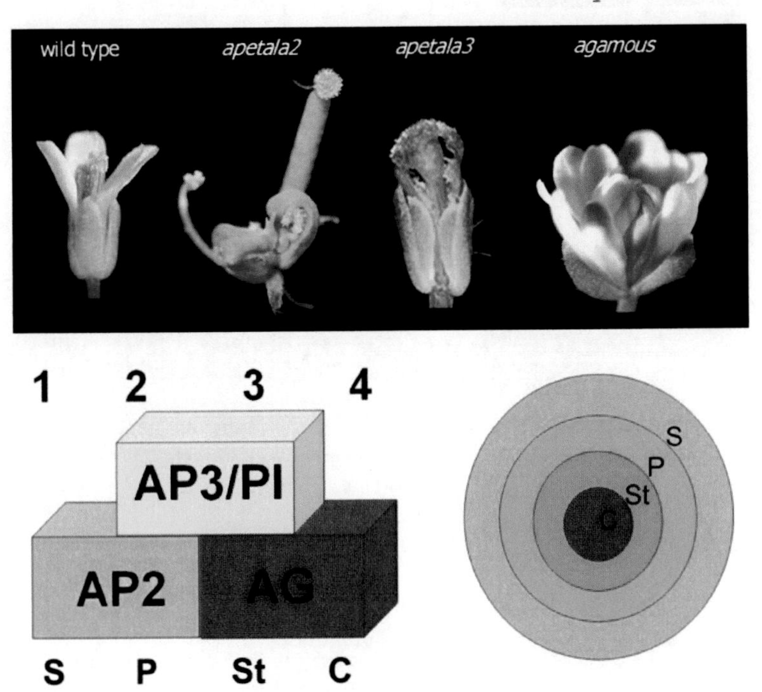

Figure 17. Structure of homeotic mutants flower and the genetic model of flower organs differentiation in *Arabidopsis thaliana.* At the bottom left – the scheme of a half of flower side view; at the top there is the number of a verticil; at the bottom there is the type of an organ to be formed in the verticil (S – sepal, P – petal, S – stamen, C – carpel, L – leaf). At the bottom right – the scheme of a flower, dorsal view; blocking of expression domains is marked with a color scale (green color of the 2nd verticil is the result of combined expression of *AP2* and *AP3/PI* genes, orange color in the 3d verticil is the result of combined expression of *AP3/PI* and *AG* genes (from Ezhova, 2010).

Genes of ABC-classes

Genes	*Antirrhinum majus*	*Arabidopsis thaliana*
A-class	*SQUAMOSA (SQUA)* *LIPLESS1 (LIP1)* *LIPLESS2 (LIP2)*	*APETALA1 (AP1)* *APETALA2 (AP2)*
B-class	*DEFICIENS (DEF)* *GLOBOSA (GLO)*	*APETALA3 (AP3)* *PISTILLATA (PI)*
C-class	*PLENA (PLE)*	*AGAMOUS (AG)*

Domenic organization of proteins coded by homeotic genes of plans

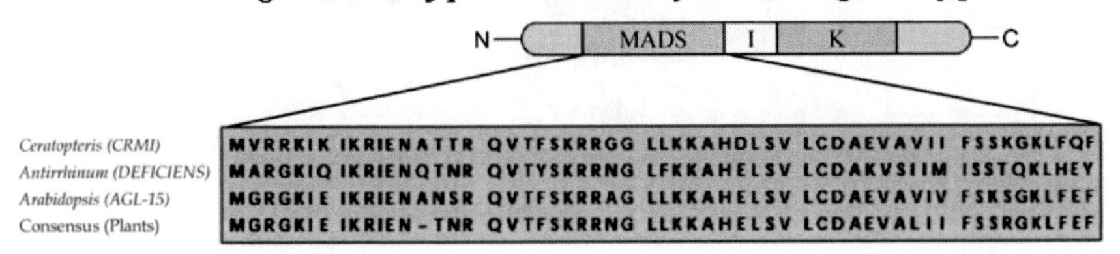

Figure 18. Domain arrangement of proteins coded by homeotic genes of plants. Apart from MADS-domain proteins contain the conservative K-domain (keratin-like domain) which provides interaction with other proteins. Domains I and C are less conservative: I can participate together with K-domain in formation of protein dimers, while C domain can take part in regulation of transcription of other genes (from Ezhova, 2010).

Seed propagation of plants leads to increasing of species quantity. The final proof of propagation intensity is considered to be the seed crop.

Vegetative Propagation

In the classical meaning, vegetative propagation is *the increasing of species individuals* in number as the result of separation *of the living available parts of the plant vegetative body* (buds, shoots, roots, etc.); in many cases, this process is accompanied by *regeneration*, i.e., *restoration* of missing organs in separated plant parts. This is essentially distinguished the vegetative propagation from the seminal one, in the course of which during *seed germination the individual of full value "is being born" from the embryo at once*.

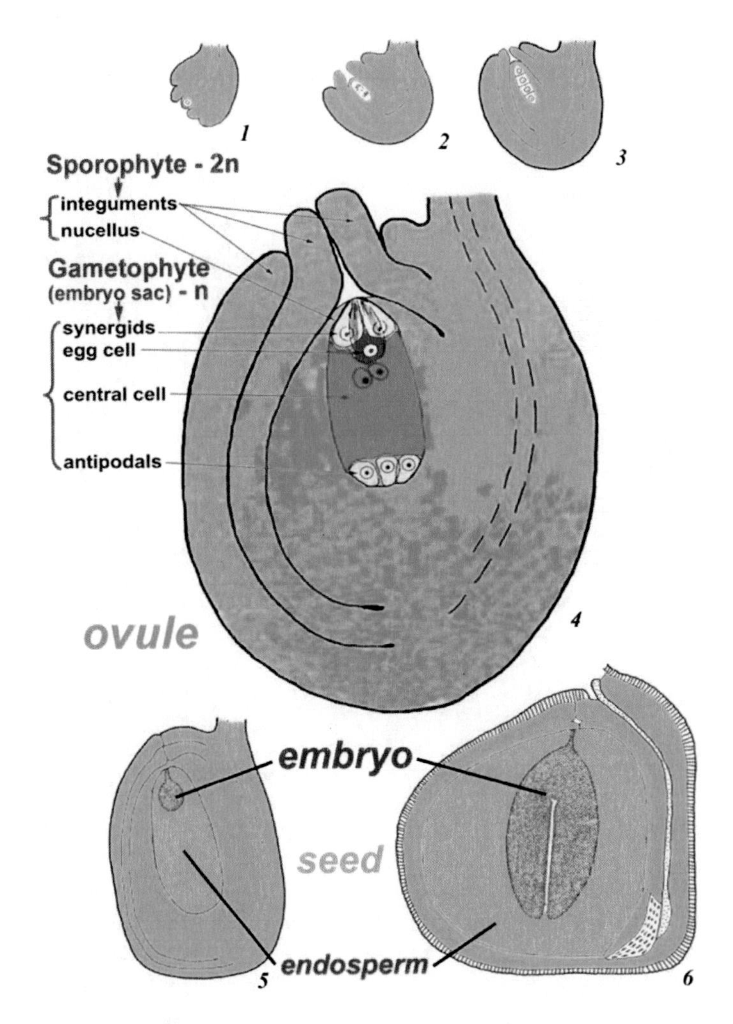

Figure 19. Polygonum-type of embryo sac development: (1-3) early stages of the ovule development – megasporogenesis; (4) ovule at the stage of mature, 7-celled embryo sac; (5, 6) seed at different stages of embryo and endosperm development. The forming vascular system is seen.

Posterity, arising as a result of vegetative propagation, independently on the place of its formation (vegetative or generative structures), produces *clones* — totality of the genetically homogenous individuals, identical by genotype to the maternal organism. According to new data on vegetative propagation, the clone sometimes is supposed to consist of not rather uniform individuals as a result of mutational, etc., processes.

Polygonum-type of gametophyte development (3D reconstruction)

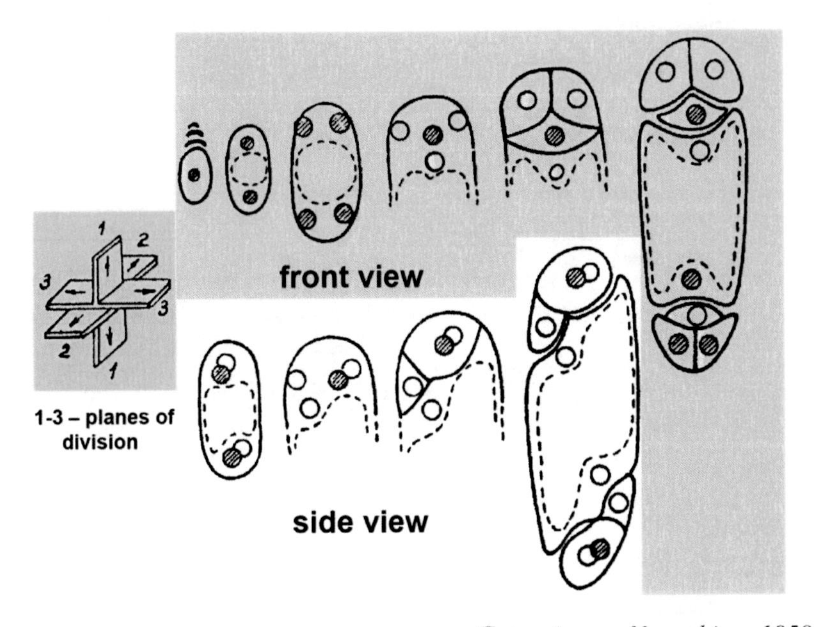

Gerassimova-Navashina, 1958

Figure 20. Polygonum-type of embryo sac development as a model of ontogenesis and evolution.

Vegetative propagation occurring *without the participation of a man* is referred to as *natural,* while the one happening with his participation is *termed artificial.* In natural conditions, the efficiency of vegetative propagation is defined by *the age* (or age condition) of maternal plant, *the longevity of its physiological contact* with vegetative posterity and by *the degree of remoteness and rejuvenation of the later.*

Certain types and modes of vegetative propagation were distinguished: *sarmentation and particulation* (Smirnova, 1974; Barykina, 2000a, b). The structural unit of sarmentation usually is a brood bud. In botanical literature the notion “sarmentation” should be used for identification of one of the natural vegetative propagation modes, at which the intensive capture of the territory observes and the highest degree of survival of daughter plants, that undergo quick rejuvenation, riches. By the depth of posterity rejuvenation such mode of vegetative propagation usually approaches to the seminal one (Levina, 1981; Barykina, 2000a).

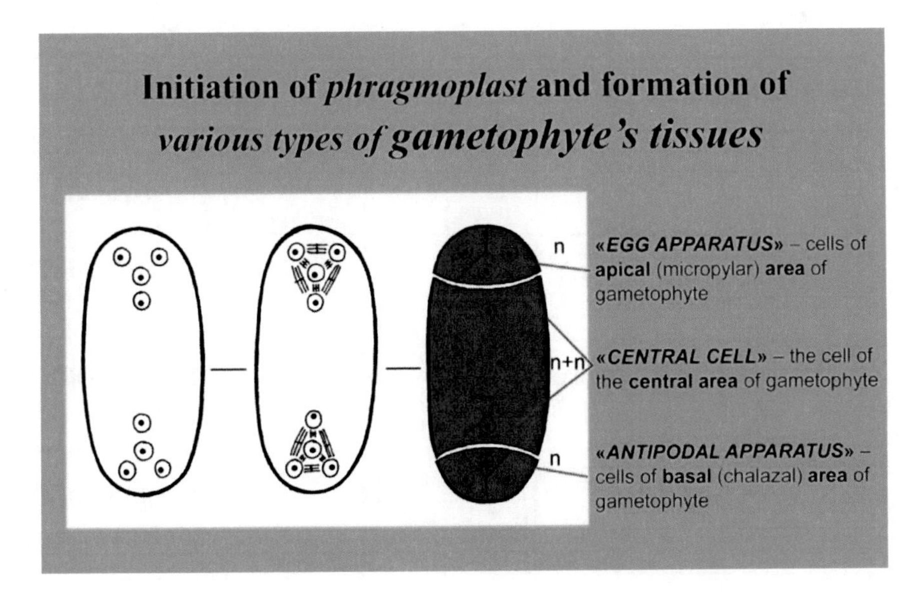

Figure 21. From Batygina, 1996b.

The bud is a *"monopolar" structure* presented by initial shortened shoot or its part (Figure 24). In a broad sense, both the dormant buds and the tops of growing shoots are called buds. In a narrow sense, only the structures that stay in rather dormancy are considered to be buds. Moreover, by the type of laying down of initials the buds are divided into *vegetative and generative (flower)* with the initials of reproductive organs (flowers and inflorescences). Besides the pure *vegetative* buds there are also *vegetative-generative* ones, in which the series of vegetative metameres are laying down and the apical cone is transferred into initial flower or inflorescence. *The bud that contains the initial of a single flower is referred to as a flower bud.* By the *position on the plant and origin the terminal, lateral and adventive buds* are distinguished. The terminal buds are formed on tops of shoots; the lateral in the leaf axils.

We established for the first time that new individuals at *vegetative propagation* in flowering plants could be produced by means not only of the bud, but the embryoids (somatic embryos) as well (Figure 2, 3, 4). In this connection, I consider it possible to express *the untraditional view on the types and modes of the seed and vegetative propagation, their ratio in the system of flowering plant reproduction, morphogenesis pathways standing in their basis and mechanisms determining the change of developmental program during ontogenesis and phylogenesis.*

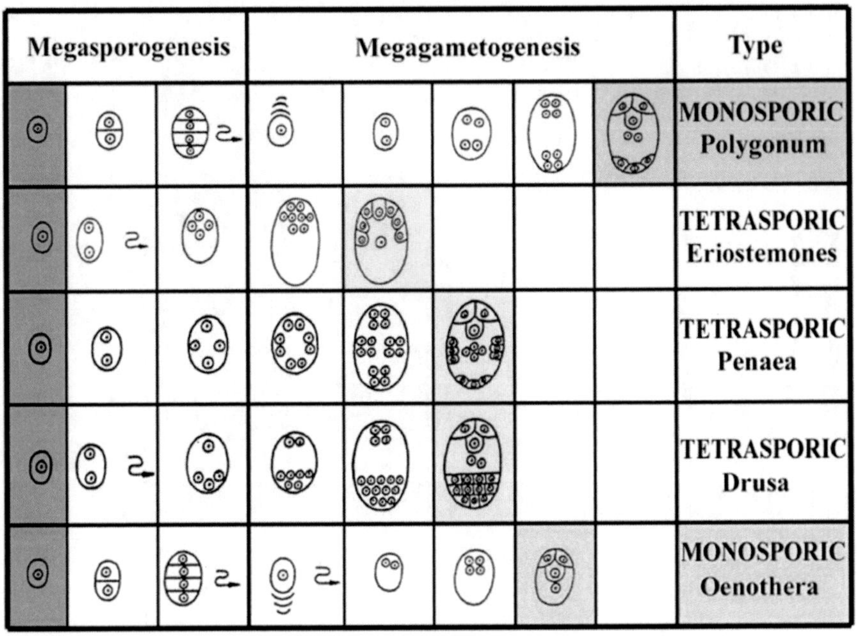

Megasporogenesis	Megagametogenesis	Type
		MONOSPORIC Polygonum
		TETRASPORIC Eriostemones
		TETRASPORIC Penaea
		TETRASPORIC Drusa
		MONOSPORIC Oenothera

Figure 22. Biodiversity of embryo sac development types.

Viviparity

Despite of many studies, a biological sense of some modes, types and forms of propagation is still unclear, as well as their relationship and the evolutionary role. A *viviparity* (lat. *vivus,* living; *pario,* to give birth) takes a special place among them.

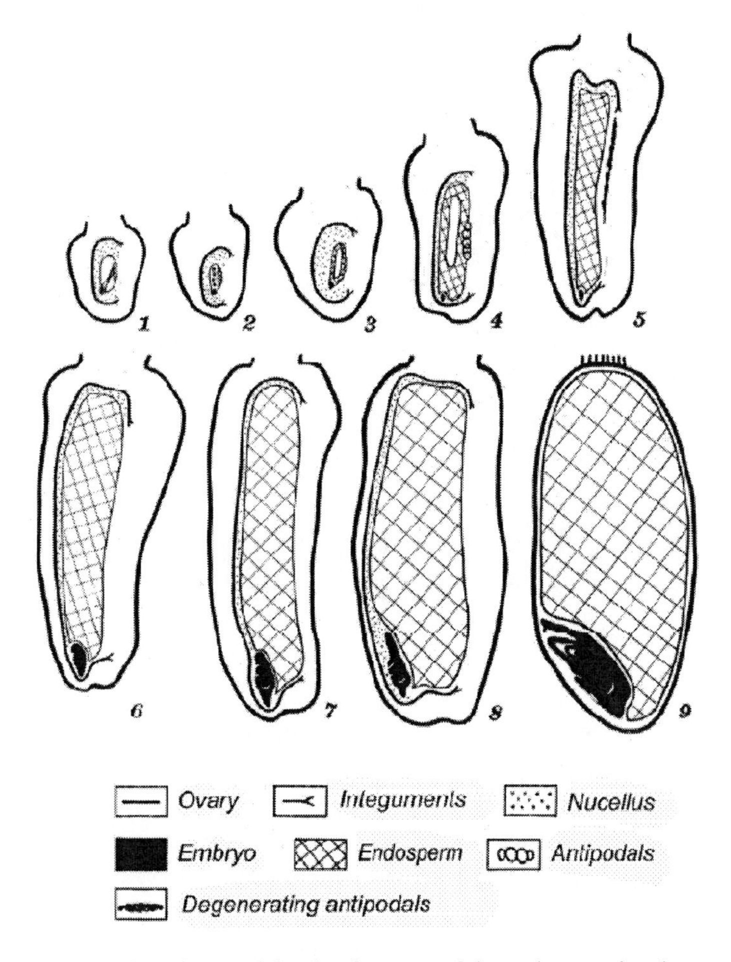

Figure 23. Successive phases of the development of the embryo and endosperm in the caryopsis of wheat, Diamant : (1) after 5–6 h; (2) 1 day; (3) 2; (4) 3–4; (5) 7–8; (6) 9–10; (7) 11–12; (8) 13–15; and (9) 25–30 days after pollination (from Batygina, 1987a).

However, the notion "*viviparity*" is still uncertain and debatable because of insufficient information available on its structural-functional bases. There is no common opinion what the term "viviparity" means: either germination of seeds on the maternal plant (in some *Rhizophora, Bruguiera, Kandelia, Ceriops* and *Avicennia*) or young plant development on the different parts (leaf, stem, inflorescence) of the mother plant (in *Poa viviparum, Polygonum viviparum, Cardamine, Bryophyllum*). However, most of the botanists *classify* a germination of mangrove seeds as well as of vegetative diaspores in other plant species *under viviparity*.

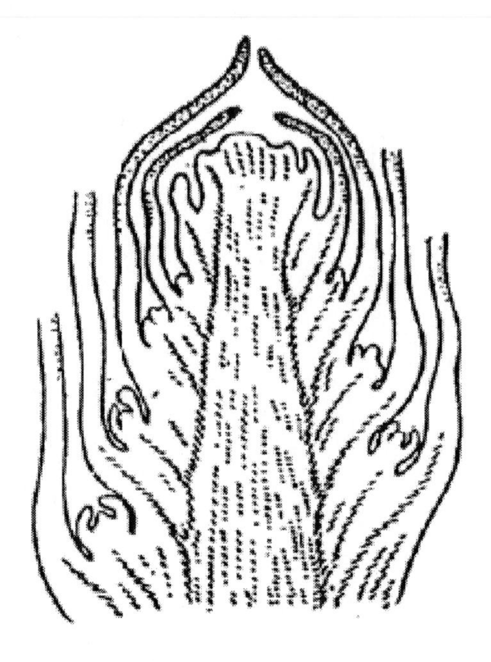

Figure 24. Longitudinal section of a bud.

Studies of viviparity are fundamentally important for research of morphogenesis, relationships between maternal and daughter organisms, transport of nutrients and hormones, their influence on the daughter organism development, formation of *its autonomy*, and reproductive strategy of species. It is also important for solving such applied tasks as a selection of optimal conditions for seed storing and their germination, cultivation of the endangered species in vitro for further repatriation.

Animal viviparity has been already known to Aristotle (fourth century BC). He used it as a character to divide the animals in viviparous and ovoviviparous. Linnaeus (1737) also applied this character classifying species from the genera *Festuca* and *Poa*.

Scientific names of some plants (*Festuca ovina var. vivipara, Polygonum viviparum, Eriogonum viviparum, Poa vivipara, Remusatia vivipara, Asplenium viviparum*), as well as animals (*Xenopleura viviara, Viviparus viviparus, Zoarces viviparus, Barbus viviparus, Lacerta vivipara, Patriella vivipara, P. paravivipara*) were created being based on the term "viviparity".

Viviparity in plants is a special type of reproduction wherein embryos in a seed, fruit or brood bud develop and form seedlings (propagules) without a dormancy period of the maternal plant. Genotype of the descendants depends on the propagule origin.

At the moment this phenomenon has been discovered in 281 species of the flowering plants belonging to 55 families (Nymphaeaceae, Ranunculaceae, Orchidaceae, Poaceae) (after Batygina et al., 2006). Besides, there are 197 species of viviparous ferns of 8 families. The viviparous species of *Bryophyllum* (Crassulaceae) is considered to be one of the most interesting objects of great practical significance.

Viviparity in animals is a type of sexual reproduction wherein either larva or juvenile develops inside the body of the maternal organism, and is released free of egg envelopes. Viviparous species are already known among ammonites.

Genetic Grounds for Viviparity

With normal development of a seed, the phase of maturation involves the synthesis of reserved nutrition substances, ceasing of the embryo's growth and development of resistance to drying. It is known that abscisic acid (ABA) is the key regulator of gene expression in late embryogenesis, which correlates with responses to various extreme conditions. The change in synthesis and sensitivity to ABA can be one of the reasons leading to viviparity.

Considering content and degree of sensitivity to ABA viviparous mutants of corn were divided into two classes. In mutants of the first class (*vp2, vp5, psl* (*vp7*), *vp8, vp9*) there can be observed the decreased level of ABA and change in synthesis of carotenoid (with exception of *vp8*).

In *vp1*, the mutant related to the second class the level of endogenous ABA in the embryo is not changed, though the mutant is not sensitive to exogenous hormone. Besides, in *vp1* mutant the synthesis of anthocyans in cells of aleurone layer is depressed. The product of *Viviparous-1* (*VP-1*) gene is included in regulatory proteins, stimulating maturation and dormancy of caryopsis. The product of *VP-1* gene is the factor of transcription and is required for the expression of *Cl, Globulin* и *Em* genes. Besides, the protein *VP1* acts as a repressor of genes of α-amilase, functioning at germination in cells of aleurone layer.

In seeds of viviparous mutants of corns, genes participating in synthesis and recognizing of ABA have been discovered. Embryos of one of mutants (*vp14*) have normal sensitivity to ABA. However content of ABA in mutant embryos was by 70% lower, than in embryos of a wild type, which point at the disorder in synthesis of ABA. Protein *VP14* catalyses the reaction of disintegration of carotenoids.

A lot of viviparous mutants of *Arabidopsis* have phenotypes similar to those of viviparous mutants of corn (for example, *abi3* mutants). Mutation of *abi3* occurs in the period of dormancy of the seed, gaining of spare proteins and fats, decay of chlorophyll, ability to respond to the action of ABA and resistance to drying. Characteristics of the existing mutant *abi3*-alleys have confirmed the hypothesis that the protein *AB13* participates in the cascade of reactions of perception and transduction of ABA. Normal perception of ABA by the *AB13* gene is a necessary but not sufficient condition for many critical stages connected with embryogenesis in *Arabidopsis*. Similar successions and analogous phenotypes of *VP1* and *AB13* prompt the idea that they are functionally homologous genes.

In *Arabidopsis* mutants *leafy cotyledon* (*lec1*) and *fus3* the precocious germination of seeds is rarely observed. Comparison of these mutants provides the ground to believe that genes *LEC1* and *FUS3* can be responsible for connected but not identical functions during embryogenesis. They code regulatory factors which activate a wide range of embryogenetic programs starting from the heart stage. *FUS3* and *LEC1* positively regulate increased content of protein *AB13* in the seed.

Viviparity is determined by the genotype of the embryo and is not dependent on the genotype of the endosperm.

It is possible, that mechanisms of regulation of viviparity in the mangrove will be similar to those of in mutants of corn and *Arabidopsis*. Probably, it will be possible to find some similarity in regulatory processes taking place at vegetative viviparity.

Chapter III

Nontraditional Notions of the Types and Modes of Reproduction

Morphogenesis Pathways

The active application of the method of *in vitro* culture in experimental and biotechnological investigations constantly excites interest to the series of unsolved problems of morphogenesis repeatedly examined in classical morphology and gives a rise certain new questions. For instance, what are the pathways and modes of morphogenesis at the formation of a new individual in the culture *in vitro* and could they be compared with morphological events occurring during the formation of new individual in natural conditions? What are the initial cells giving embryoids in the culture *in vitro* and in natural conditions characterized by? Why do the isolated somatic cells in one taxon produce the buds, in others the roots and in some the embryoids? Is there parallelism in the development of sexual embryo and embryoid, forming in natural conditions and in the culture *in vitro*?

The question on the modes and mechanisms of switching over developmental program from one morphogenesis pathway to another in the process of ontogenesis and during the life cycle is considered to be one of the actual.

In the 1970s, contradictory opinions concerning morphogenesis pathways in the culture *in vitro* existed. According to the generally accepted view, the formation of new individual in flowering plants realizes either by sexual mode

— the sexual process preceding embryogenesis (heterophasic reproduction) or by asexual (vegetative) mode—the formation of buds and roots (homophasic reproduction), i.e., regeneration occurs (Battaglia, 1963; Batygina, 1999a-c). However it is appeared that upon the homophasic reproduction new individual creates not only by regeneration (organogenesis—gemmorhizogenesis), but also by means of embryoids formation — embryoidogenesis.

In the light of data on flowering plant reproduction *in situ, in vivo* and *in vitro* we have revealed for the first time that *the vegetative propagation* is presented by two types: *embryoidogenous* (embryoids) and *gemmorhizogenous* (bud plus roots) but *not only a single gemmorhizogenous, as earlier considered.* At the formation of new individual not two (embryogenesis and organogenesis, as earlier considered) but three pathways of morphogenesis exist: *embryogenesis, embryoidogenesis* and *organogenesis.* The pathways and modes of new individual formation given are universal both in natural and experimental conditions (Batygina 1984) (Figure 1, 2, 3). It had allowed us distinguish a new category of vegetative propagation – embryoidogeny and examine the role of this phenomenon in the system of flowering plant reproduction. Long standing elaboration of theoretical bases of plant reproduction permits to *establish the status and interaction of different types, modes and forms of seed and vegetative propagation*, realization of each in a rather extent determines the *reproductive strategy of the species* (Figure 1). The creation of new classification of types and modes of flowering plants reproduction and propagation considers being a result of this elaboration (Batygina, 1987a, b, 1989a, b, 1990a, b, 1991a, 1992a, b, 1993a, 1994a-c, 1996a, 1997, 2000, 2005a, b, 2006a, 2009a; Batygina Vasilyeva, 2002; Batygina et al., 1978).

The theoretical elaborations in the sphere of embryoidogeny already have permitted to obtain the certain plant forms and sorts, including that on the basis of androcliny phenomenon (see Batygina, 2005b; Batygina et al., 2007, 2010).

Embryoidogeny: A New Category of Vegetative Propagation as a Basis for Changing of Morphogenetic Developmental Programs

Embryoidogeny (Greek *embryon,* embryo; *oidos,* view; *genus,* origin) is one of the two types of homophasic reproduction of flowering plants *in situ, in vivo* and *in vitro,* the elementary structural unit of which is an embryoid. *Embryoid* is the incipient individual forming asexually *in situ, in vivo* and *in vitro*. Synonyms are *somatic embryo* and *embryo-like structure*.

The main thesis of the concept of embryoidogeny is the belief in the universality of the morphogenesis of embryoids as well as sexual embryos, developing in natural conditions and in experimental ones, in in vitro culture (Batygina, 1996a, b, 1998, 2006a, b, 2009; Batygina, Zakharova, 1997a, b).

Embryoid, like embryo, is characterized by the formation of its own *new axis* (on relation to the maternal organism), joining polar developing shoot and root apices. As a rule embryoid does not obtain the vascular system common with maternal organism (close radicular pole). Genesis of embryoid, as well as its shape and sizes are taxon specific. *The main features established for sexual embryos (polarity, cellular and histogenic differentiation, autonomy, etc.) are typical for embryoid* also. During embryoid germination as at embryo germination a new individual just "births" (Batygina, 1987b, 1989a, b, 1993a, b, 1997, 2000, 2006a, b, 2009a).

Together with this, *embryoid possesses the principal (by the origin) resemblance with the bud,* that is firstly in *the mode of formation – homophasic reproduction* (sporophyte → sporophyte), that *unites them as elementary structural units of vegetative propagation.*

The main difference of embryoid from the bud is that *the latter from the very beginning* of its development represents only *the part of a whole organism – "monopolar" structure* (*has only shoot apex and the open raducular pole, the cotyledons are not formed*), connected with the maternal plant by means of vascular system (Figure 4). Just in the process of further development its connection with maternal organism interrupts and as far as the root (roots) develops it becomes bipolar. However in embryoids of some species as in the buds the short-time connection with the maternal organism is observed; moreover, the longevity of contact and the stage of embryoid development on which this contact is realized are differed in various plant species (see below). *It allows speaking about the existence of transitional, intermediate forms between embryoid and bud.* The research of such intermediate forms and the revealing of their *structural-functional characteristics* have a great significance regarding their utilization *in the work on cloning plants in biotechnological practices* (for example, the definition of monopolarity or bipolarity of above mentioned structures).

Thus, while *distinguishing the embryoidogeny into the special type of reproduction and propagation, we used two criteria: ontogenetic (homophasic reproduction, accompanied by neither meiosis nor the process of gamete fusion, i.e., asexual mode of new generation formation; uniparental heredity) and morphological (bipolar organization of the structure with shoot and root apices and new polar axis). As to embryogeny (heterophasic reproduction), the formation of new individual occurs to be a result of sexual process, i.e., meiosis and gamete fusion (biparental heredity).*

The tendency of the embryoid formation realizes *at all stages of flowering plant ontogenesis, beginning with the zygote*. Embryoids can arise on *the different structures and organs of plants exogenously or endogenously*, usually from one somatic cell, rare from embryonal cellular complex. The phenomenon of embryoidogeny is observed in species, growing in various ecological zones.

Depending on the origin and location of embryoids on maternal plant, *two main forms* of embryoidogeny can be distinguished: *floral* (reproductive) and *vegetative* (Figure 1).

Floral (reproductive) *embryoidogeny* is the formation of embryoids in the flower and seed structures. It comprises in its turn embryonal[1] monozygotic "cleavage" (*monozygotic* splitting) on the basis of zygote or embryo; ovular, from integumentary and nucellar cells in ovule; and also gametophytic embryoidogeny, from the cells of micro- and megaspores, male and female gametophytes.

It is interesting to examine the phenomenon of *embryonal monozygotic embryoidogeny*, the formation of monozygotic twins, as one of the modes of homophasic reproduction.

The formation of monozygotic twins is the result of producing of one or more embryoids from a single initial cell, zygote or the cell of multicellular embryo, by means of different modes, that is likely conditioned by the character of division (equal, unequal, etc.), the time and the place of the

[1] Introduction of the terms "embryonal", "ovular" and "gametophytic" embryoidogeny as well as "foliar", "cauligenous", "rhizogenous" embryoidogeny (see later) permits correct identification not only of embryoid origin (from somatic cells of embryo or ovule), but also the embryogenesis stages at which they form.

formation of these twins. The question about the stage, when the embryoids form, and what is the mechanism of their formation, often remains to be disputable. Per se, it is considered to be vegetative propagation and the cloning of the organism at different ontogenesis stages.

Embryonal Monozygotic Embryoidogeny

The term ***embryonal monozygotic embryoidogeny*** emphasizes that the embryoids arise from the somatic cells of the sexual embryos at different stages of their development, as a rule at initial ones. For instance, upon the embryoid formation from suspensor cells a single zygote gives rise to the two monozygotic twins, one of which occurs to be a sexual embryo, another one an embryoid. At equal zygote division, the both monozygotic twins derive by asexual mode of individual formation, and vegetative propagation, i.e., such twins by their origin are considered to be embryoids, probably, of the same genetics because of meiosis process lack before their formation.

The notion "monozygotic twins" occurs to be rather capacious and comprehends all cases of the formation of genetically identical clones from a single zygote at different stages of embryogenesis. From the other hand, the notion of *monozygotic embryoidogeny* is likely to be considered not only in a broad sense (as a synonym of the monozygotic twins formation), but also in a narrow sense, limiting by the cases when the clones are derived immediately by the account of zygote division (splitting).

The phenomenon of monozygotic embryoidogeny was observed in different members of flowering plants: *Loranthus* (Loranthaceae), *Erythronium, Tulipa* (Liliaceae), *Limnocharis* (Orchidaceae) (Ernst, 1918), *Lobelia* (Lobeliaceae) (Crété, 1938), Papaveraceae (Ilyina, 1968), Poaceae (Erdelská, 1996) and in many other species of flowering plants.

Jeffrey (1895) in details described this phenomenon in *Erythronium americanum*. After fertilization the zygote divides with the formation of irregular group of cells ("embryogenic mass") at the lower end of which the protrusions develop (as a rule in 2–3 or 4 in number) becoming independent embryos (embryoids).

In *Limnocharis emarginata* (Hall, 1902) zygote divides transverse into two cells, thereby the largest basal cell in certain cases is able to produce embryogenic mass of cells developing similar to that in *Erythronium*.

In the members of the Orchidaceae family (*Eulophia epidendrae,* Swamy, 1943) *several somatic embryos* (embryoids, T.B.), *derived from the somatic cells of the sexual embryo could arise in the same embryo sac* (Figure 25). According to the author, "budding" is possible. The figures he gave and the discussion of data Fagerlind fulfilled (1946) can demonstrate that "the bud" and "protrusions" could be considered as embryoids.

Monozygotic embryoidogeny

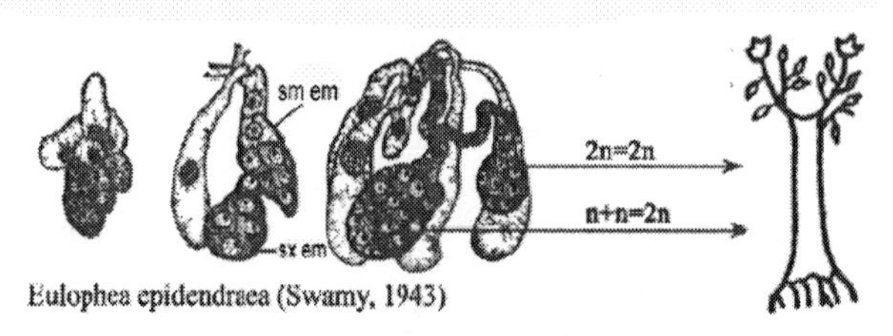

Nucellar embryoidogeny

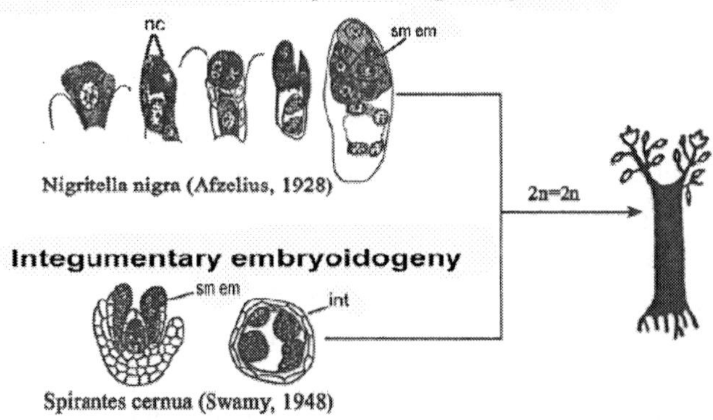

Gametophytic apomixis (apogamety)

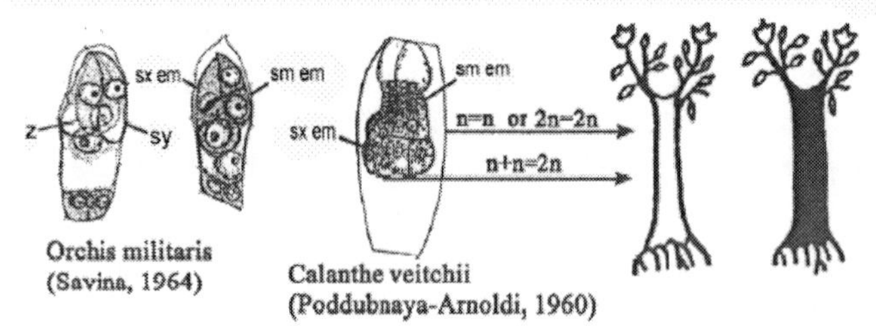

Figure 25. Pathways of formation of the embryoid and polyembryony in orchids: *z* – zygote; *int* – integument; *nc* – nucellus; *sx em* – sexual embryo; *sy* – synergid; *sm em* – embryoid; ■ – uniparental inheritance; □ – biparental inheritance (from Batygina, 2000).

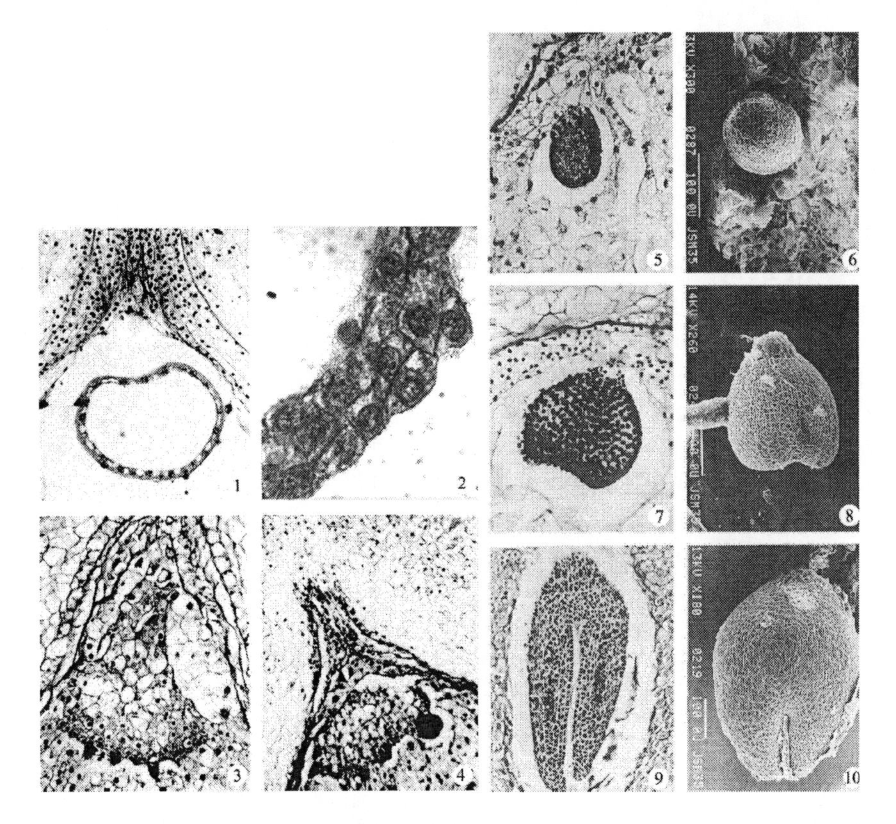

Figure 26. Monozygotic embryoidogeny in representatives of genus *Paeonia*: (1) multi-nuclear coenocyte; (2–4) coenocyte-cellular structure with embryoids; (5–10) globular (5, 6), heart-shaped (7, 8) and torpedo-shaped (9, 10) stages of the development of one of embryoids at coenocyte-cellular proembryo (the others degenerate), endosperm cellular, inner integument is practically destroyed; (1–5, 7, 9) LM and (6, 8, 10) SEM (from Brukhin, Batygina, 1994).

In *Eulophia epidendrae*, three different varieties of the appearance of such embryoids were discovered:

- after fertilization, the zygote divides with the formation of embryogenic mass, from the chalazal cells of which numerous embryoids arise;
- the zygote results in filiform proembryo that "branches", and the branches transform into a new embryoids;
- shoots that "bud" from the proembryo develop into new embryoids.

The special type of “cleavage” embryoidogeny is observed during embryo development in the members of the *Paeonia* genus (Figure 26).

Peony ontogenesis is the unique model for study of developmental programs and their switching during the life cycle.

The formation of sexual embryo in all representatives of peony is considered to be unique as at the first stage of its development the coenocyte produces from zygote without cell wall formation (Yakovlev, Yoffe, 1961). At the later period of sexual embryo development (heterophasic reproduction) numerous embryo-like structures produce at the coenocytic-cellular and cellular stages, *one of which modifies into the sexual embryo*, according to Yakovlev (1983) concept. *According to our notions, these structures represent the embryoids, arising from epidermis cells of sexual embryo.* This occurs to be one of the forms of monozygotic embryoidogeny (Batygina, 1987b, 1989a, b, 1992a, b; Brukhin, Batygina, 1994).

So, *in the mature seed, the only embryoid* remains as a result of competing development; the rest die (*apoptosis*) at the different developmental stages. *The sexual embryo also degenerates*; the embryoids produce on it, herewith its destruction takes place at the different stages of seed development in various peony species. In this case *the cloning of organism happens at the first stages of its development* (embryo) and in the mature peony seed there is, as a rule, *a single embryoid* (somatic embryo), represented the *clone* of *sexual embryo* but not *a proper sexual embryo*, as it was expected earlier (Yakovlev, Yoffe, 1961; Yakovlev, 1983). This is believed to be a striking example of *transition from one mode of sporophyte formation—sexual (n+n)—to another—asexual (embryoidogenesis—2n=2n)*. In the process of *evolution* the representatives of one taxon obtained the ability to homophase reproduction (*sporophyte→ sporophyte*) at the very early ontogenesis stages, owing to that the vegetative propagation by cloning takes place (uniparental heredity) (Figure 27, 28).

The detailed cyto-embryological investigations revealed in the developing peony seed *two types of successively forming embryos, differing in origin: sexual embryo with non-differentiated organs, which stops developing and is replaced by the formation of another, somatic embryo, differentiated into all general organs* (shoot and root apices, cotyledons).

While examining the reproduction in different peony representatives, its important peculiarity should be mentioned. It is in the fact that the *vegetative propagation*, presented by two types, *embryoidogenesis* and *gemmorhizogenesis*, becomes the *main mode of peony reproduction*. At *embryoidogenesis* the formation of embryoid takes place in the seed, the epidermal stem cell of sexual embryo occurs to be the initial of it. At *gemmorhizogenesis*, playing the general role during the formation of peony shrub, the laying down of monopolar structures, the *buds* and the *shoots* on the *rhizomes* occur. I would like focus attention to the fact that in the process of bud laying down during the formation of both shrub and flowers *the stem cells of apex play a significant role* as well as at embryoidogenesis. Thereby it should be noted once again that two reproductive modes in the peony represent per se *vegetative propagation.* Although in the peony the sexual process, gamete fusion resulting in zygote producing, occurs, however owing to "falling down" of the final stages of sexual embryo development, the sexual mode of individual formation in general sense is absent.

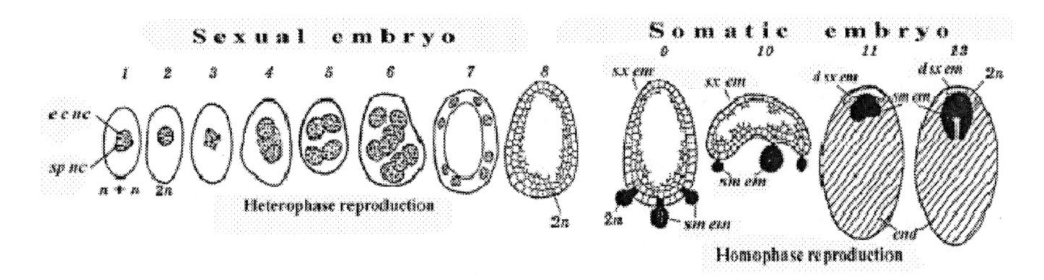

Figure 27. Switching over the developmental program from the heterophasic to homophasic reproduction in *Paeonia* seed: (1–8) development of the sexual embryo from the zygote to coenocyte-cellular stage (heterophasic stage of reproduction): (1) pollination; (2) zygote; (3–7) development of coenocyte embryo; (8) formation of epidermis at coenocyte-cellular stage; (9–12) formation of embryoids and degeneration of the sexual embryo (homophasic stage of reproduction): (9, 10) several embryoids at sexual embryo; (11) one embryoid at heart-shaped stage, remains of the sexual embryo can be seen; (12) mature seed with embryoid;

d sx em – degenerating sexual embryo; *sx em* – sexual embryo; *sm em* – embryoid; *end* – endosperm; *sp nc* – sperm cell nucleus; *e c nc* – egg cell nucleus (from Batygina, 1997).

All of the factors mentioned above allow taking into consideration the importance of reproduction aspects listed above in the research of the peony population structure and coenosis formation with its participation.

Multistage morphogenetic program of peony life cycle is related with activity of stem cells, cloning and terminal differentiation at different hierarchy levels.

Some experimental data concerning the influence of different factors (radiation, morphactins, morphoregulators, etc.) on the development of sexual embryo of *Eranthis* and other plants in culture *in vitro* (Haccius, 1965, 1978, Figure 29) are comparable to the phenomenon described for peony, i.e. with changing of morphogenetic developmental program from sexual to asexual one (Batygina, 1992a, b).

It permits to assume that the stress situations in nature likely induced transition from sexual reproduction to asexual one during the evolution course.

In some cases (*Paeonia*) in nature this process turned to be fixed in the course of evolution; in others (*Eranthis*), the realization of embryoidogenic pathway of morphogenesis fulfils sporadically, only in stress situations.

In all cases of morphogenetic changes of the *complete program* of life cycle both in natural conditions (*Paeonia*) and in culture *in vitro* (*Eranthis*), stem cells determine the mechanism of critical stages in the ontogenesis of different angiosperms, which is taxon specific.

A special case of monozygotic embryoidogeny is the embryogenesis of coniferous and other gymnosperms (Figure 30, 31). After fertilization in the zygote of gymnosperms, several free-nuclear divisions (from 2 up to 8) usually occur. Many investigators refer the coenocytic zygote to as a free-nuclear proembryo (Dogra, 1984, a.o.). After divisions and direct transitions, nuclei complete the process of cell formation on the basis of secondary spindle, resulting in the formation of the primary cellular proembryo. After this, as a result of cell divisions, *the secondary proembryo is created in which longitudinal splitting or, correctly, dividing into cells (cleavage) takes place, realized owing to the differences in the rate of their growth.* The cleavage in coniferous embryos is the regular process, leading to the formation of separate embryos from the "embryonal units" on the each end of the suspensor "tube". Regarding the remaining terminology of Dogra, on the whole, for the description of coniferous embryology, E. S. Teryokhin (1996) suggests substituting the terms "primary embryo" and "secondary embryo" for the terms "primary zygote" and "secondary zygote" accordingly. In confirmation to this, he gave the examples of the cleavage in *Actinostrobus pyramidalis* and *Callitris rhomboidea,* whereby it is realized just at the transition from free-nuclear coenocytic primary zygote to cytokinesis, or in *Ephedra*, in which the transformation of "the primary zygote" into "the secondary zygotes" is fulfilled by another means but also at the unicellular state of the former.

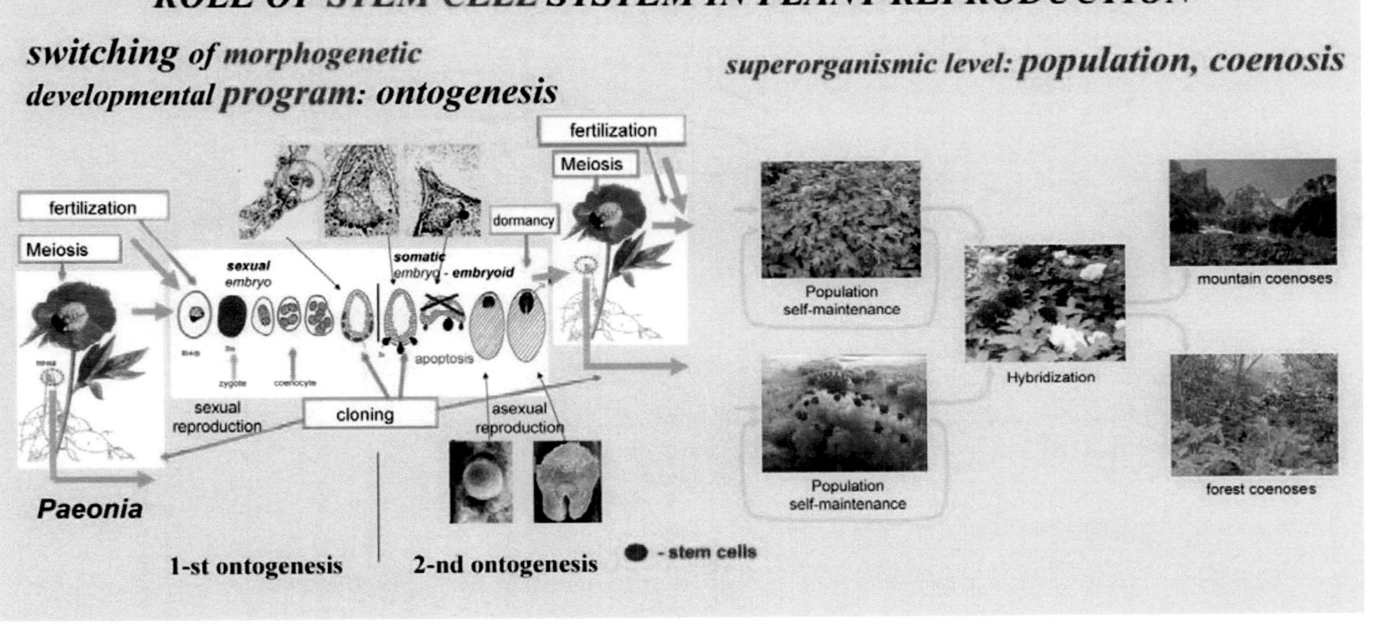

Figure 28. Role of stem cell system in plant reproduction.

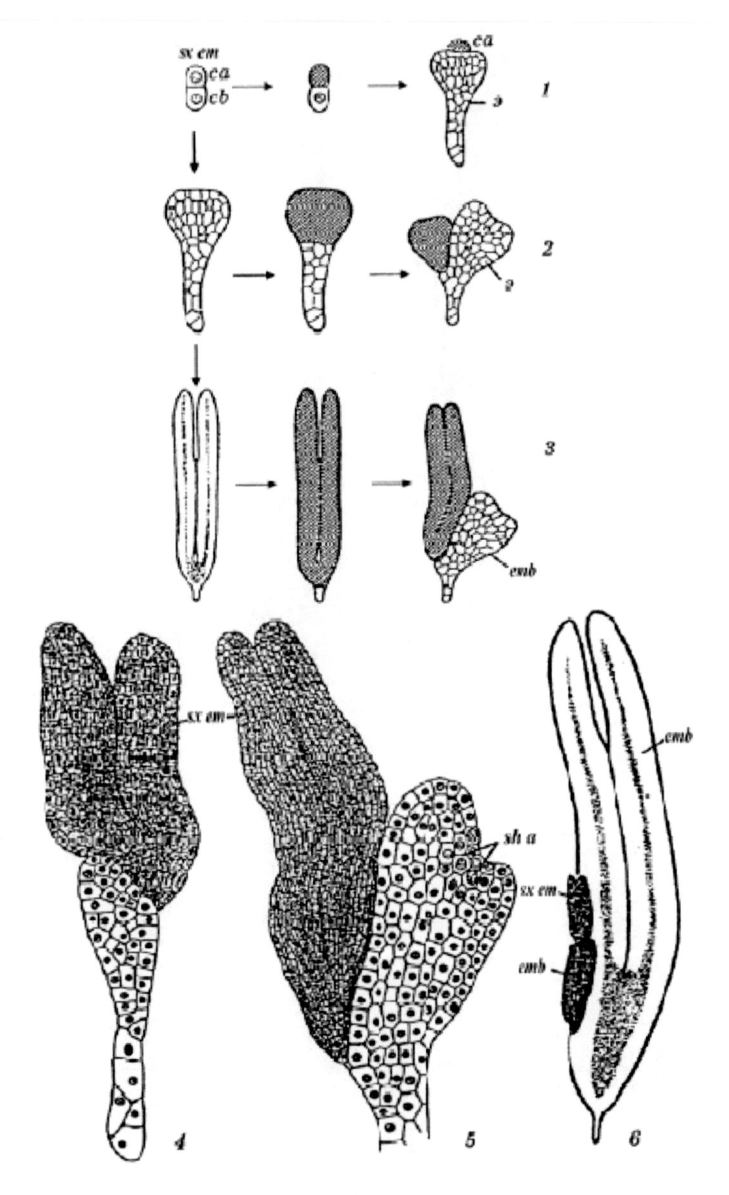

Figure 29. Formation of the embryoid from suspensor cells of the sexual embryo in *Eranthis hiemalis* in culture *in vitro*:

sh a – shoot apex; *sx em* – sexual embryo; *emb* – embryoid (from Haccius, 1965).

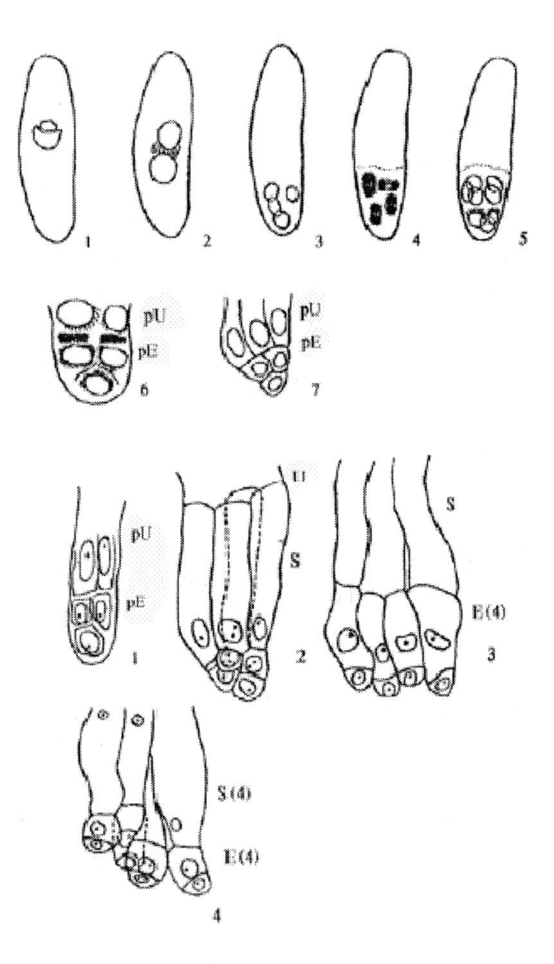

Figure 30. Two-stage embrygenesis of the gymnosperms: A. General scheme of embryogenesis in Cupressaceae and Taxodiaceae families (from Dogra, 1984): (1) pollination; (2–5) free-nuclear development of proembryo; (6) formation of cell walls; (7) cellular proembryo; B. Embryogenesis in *Thuja orientalis* (from Singh, Oberoi, 1962): (1) primary proembryo; (2) formation of the suspensor; (3) secondary proembryo, cleavage; (4) development of secondary embryos, formed as a result of cleavage; pU – primary upper (open) tier; pE – primary embryonic group; U – upper and open tier at later stages of development; S – suspensor tier of secondary proembryos; E – embryonic group of secondary proembryos (the terminology according to Dogra, 1984) (from Teryokhin, 1996).

Thus, two-phase way of embryogenesis (formation of "primary" and "secondary" zygotes) and the process of cleavage, i.e., separation of secondary zygotes and proembryo leading to polyembryony and embryo competition are considered to be the important peculiarity of embryogenesis in most taxa of the present gymnosperms (Teryokhin, 1996). In this connection it should be mentioned that the species with cleavage type of polyembryony in gymnosperms believe to be ancestors with reference to the taxa with unexpressed polyembryony (Johansen, 1950). Two phases of gymnosperm embryogenesis, as Teryokhin proposed (1991), allow considering zygote of angiosperms to be the *homolog* only to the "primary" zygotes of gymnosperms and the *embryos of angiosperms are not essentially homologous to the secondary embryos of gymnosperms* except *Gnetum*.

Speaking about monozygotic and polyzygotic embryoidogeny, the *phenomenon of fused "Siamese" embryos in vitro should be examined.* So under the *influence of inhibitors of polar auxin transport* (NPA, quercetin) on the isolated *sexual embryos of wheat at the globular stage* or at the early stage of transition to *organogenesis*, two phenotypes of "Siamese embryos" form "back to back" with numerous scutellums, turned to each other by dorsal sides and situated terminally in relation to the apical shoot meristems; "heart-shaped Siamese embryos" with the scutellums turned to each other at the ventral sides and situated laterally in relation to the apical shoot meristems (Fischer et al., 1997). Similar phenotypes with numerous scutellums and apices were observed in embryoids forming in the culture *in vitro* of anthers in spring soft wheat at higher concentration of synthetic auxin 2,4-D (Seldimirova, Titova, 2007). These data demonstrate the *generality of morphogenetic process* laying in the base of *embryogenesis in seed and embryogenesis in vitro* and general mechanisms resulting in the both cases in monozygotic embryoidogeny.

The phenomena similar to monozygotic embryoidogenesis in plants and leading to the polyembryony were noted for the embryogenesis of animal and human (Kanaev, 1968), when the initiating of monozygotic twins appeared to occur at different stages — from the stage of two blastomeres until the stage of gastrulation, i.e., that stage when initiation of intestines happens. In most animals it is a rare phenomenon; however, in some mammalia (two species of armadillo) at the rather late developmental stage normally several embryos arise as a rule instead of a single one.

Thus, both in plants and in animals at the formation of monozygotic twins, the cloning of zygote and embryo takes place.

Monozygotic and polyzygotic twins are a fine illustration of the displaying of the *polyembryony* phenomenon. In the case of seed plants, polyembryony (polyembryoidogeny, T.B.) in a broad sense (sensu lato) is the formation of several embryos both in a single seed and on the separate structures of a plant (flower, leaf, stem and root). In animals (for example in insects, mammalia etc.), analogous phenomenon manifests in the formation of several embryos in reproductive organs. This phenomenon is considered to be universal and occurs both in plant and animal organisms.

Because of the debatability of some aspects of the extremely difficult problem in developmental biology (polyembryony, embryoidogeny) concerning the time of formation, trigger mechanisms, a change of developmental programs, *mono-, polyzygotic and Siamese twins* both in *animals* (human) and *plants* (*in situ* and *in vitro*) is likely required in the *discussing of certain notions used by different investigators.*

The analysis of literature has revealed that usage of the term "cleavage" in different languages—Russian, *кливаж*; English, *cleavage*; French, *clivage*—infers different meanings, and the various dictionaries give various semantics of these terms. Probably, this is the reason for ambiguity of the determination of this notion. For instance, during investigating of the first ontogenesis stages in animals the notion "splitting" uses the most often, in western literature the "cleavage stage" is distinguished in the embryo development characterized by the fact that the larger zygote and its derivatives undergo a series of cell divisions without compensator growth of the cells formed. Russian zoologists as a rule do not use the term "cleavage stage", because they apply to this phenomenon the Russian term "splitting". However in rare cases this term is also used in plants (Yakovlev et al., 1957). The notion "cleavage" concerning the plants is used for the characteristics of the phenomenon of "splitting", "disintegration" of zygote and embryo wall. These notions need to be corrected with special accent that these terms could be simultaneously used both for plants and animals. Perhaps, the lack of nuances in behaviour of certain structures at embryogenesis, in particular the structure of walls at different stages of cell development, requires fixed attention and correction in their using.

In conclusion, I would like to join the opinion of famous embryologists Brien (1956) and O.M. Ivanova-Kazas (1977) working a lot with the problem of the asexual propagation in animals, who consider that all facts on the asexual propagation given often brings the dissonance in the concert of classical biology because they do not always conform with generally accepted views. Our nontraditional approach to the study of vegetative reproduction in plants and the discovery of embryoidogeny phenomenon and a new category of vegetative propagation in plants is a splendid illustration of this state.

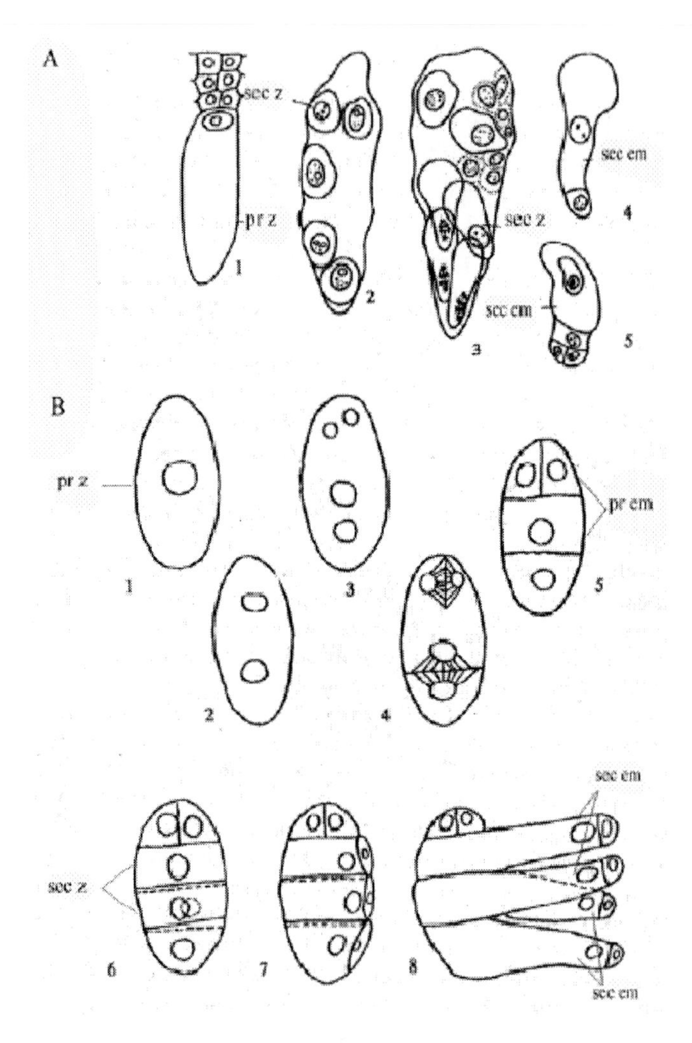

Figure 31. Different types of two-stage embryogenesis in gymnospermous plants: (A) two-stage embryogenesis in *Ephedra* with original (coenocyte) type of monozygotic polyembryony: (1) primary zygote in *Ephedra trifurcata* (from Land, 1904); (2–5) formation of secondary zygotes and development of secondary embryos in *E. foliata* (from Khan, 1943); (B) two-stage embryogenesis in *Actinostrobus piramidalis* with cleavage type of monozygotic polyembryony: (1) primary zygote; (2–4) coenocyte zygote; (5, 6) formation of secondary zygotes; (7) cleavage; (8) development of secondary embryos;

sec em – secondary embryos, sec z – secondary zygote, pr em – primary embryo, pr z – primary zygote (from Saxton, 1913 and Teryokhin, 1996).

Nucellar and Integumentary Embryoidogeny

Nucellar and integumentary embryoidogeny – the formation of embryoids from the cells of nucellus and integuments—ovule covers – was noted in more than 250 plant species. There is large information on nucellar embryoidogeny in fruit trees, such as *Citrus, Mangifera, Eugenia,* etc. The number of polyembryonic seeds can vary in different species. Thus, in *Citrus microcarpa,* the seeds usually produce 21 seedlings, whereas in *C. unshiu,* about 40.

Nucellar and integumentary embryoidogeny was also observed in some orchid species (*Nigritella nigra, Spiranthes cernua*) at different stages of ovule development, often just before embryo sac formation (Afzelius, 1928; Swamy, 1948).

Embryoids are able to arise by different modes: from somatic cells of nucellus and integuments and at different stages of ovule development (Modilewsky, 1931; Button et al., 1974; Batygina, Mametyeva, 1979; Batygina, Freiberg, 1979; Batygina, 1991a, b). Thus, in *Poa pratensis* the embryoids are formed either from a single cell of nucellus (dormant meristem), or through embryonal cellular complex as a result of proliferation of nucellus cells (Figure 32). Independently on the mode of embryoid formation, in the mature seed at nucellar and integumentary embryoidogeny one is dealing with the clone of maternal plant presented by embryoids. However, in this *Poa* genus together with agamospermous seeds, the seeds with sexual and nucellar embryos are observed.

Nucellar embryoidogeny is given attention because of various causes. One of them is the capacity of obtaining of plants without viruses. At usual vegetative propagation the cuttings are known often to be infected by different pathogens whereas nucellar embryos and seedlings planted from them are lack viruses. Some authors explain this by that the nucellus and neighbored tissues do not contact with each other by vascular system.

Geneticists and plant breeders use the high capacity of some plant species (*Mangifera, Citrus*) to form nucellar embryos as the main criterion of genotype selection for cross-breeding, ensuring thereby the diversity in generation. According to the data of many authors, nucellar embryoids derivate “rejuvenated” seedlings, which resemble ones that have arisen from sexual embryos. Maheshwary (1950) emphasized that seedlings of *Citrus,* forming *from vegetative buds on sporophyte, develop without spines*, whereas seedlings from *nucellar embryos develop into the plants with spines*. Clones of *Citrus*, always being vegetatively propagated (by cuttings), finally become weaken and sterile (Cook, 1938; Hodgson, Cameron, 1938; Frost, 1938). It is remarkable that gardeners renew the *Citrus* clones using *nucellar seedlings, developing better then plants, obtained from cuttings*.

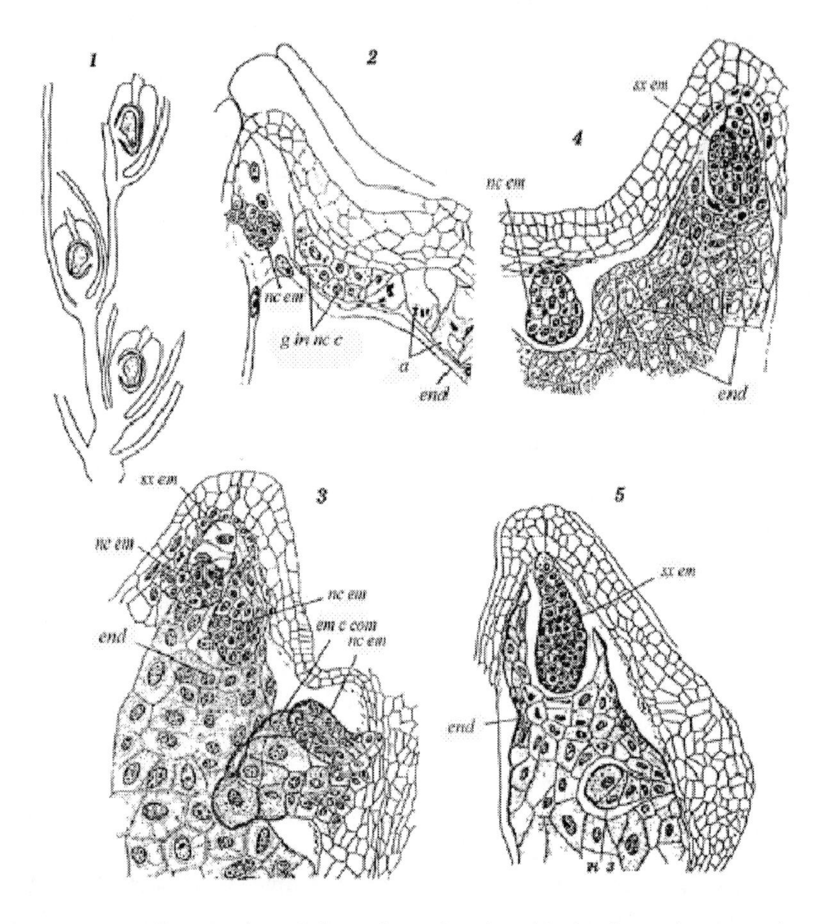

Figure 32. Various modes of formation of embryoids in the embryo sac in *Poa pratensis:* (1) spikelet; (2) micropylar part of the embryo sac with nucellar embryoid and group of initial nucellar cells, out of which embryoids are formed; (3) micropylar part of the embryo sac, sexual and nucellar embryos can be seen, embryonic cellular complex; (4, 5) sexual embryos at different stages of development and nucellar embryos, being formed out of singular cells of dormant meristem;

a – antipodals, *g in nc c* – group of initial nucellar cells, *nc em* – nucellar embryo, *sx em* – sexual embryo, *em c com* – embryonic cellular complex, *end* – endosperm (from Batygina, Freiberg, 1979).

There are cases in which, *in the same seed, the formation of embryoids precedes parallel to embryogenesis* (e.g., *Citrus, Poa*). We use the term "clone" for embryoids and seedlings arisen from them. It should be noted that *in some representatives of angiosperms* (Citroideae, Orchidaceae, etc.), *two different clones—two generations—can form in the same seed*: the clone of maternal organism (by means of nucellar and integumentary embryoidogeny) and the clone of daughter individual (*by means of cleavage monozygotic embryoidogeny*).

In that case when in the same seed there are *embryoids of different* origin: *maternal (ovular) and new daughter (embryonal)*, the plants of ***two genotypes*** *develop in population.* It is the reason of the appearance of genetic heterogeneity of seeds which together with vegetative embryoidogeny finally conditions the ***genetic heterogeneity*** *of populations*.

Gametophytic Embryoidogeny

According to our theoretical ideas, under certain stress situations in nature, in certain plant species, for example, in some of *Orchidaceae*, *the anther and the ovule with all of their structural elements* are considered to be utilized as *the units of propagation and dispersion.* It is confirmed by a great number of data on *gametophytic apomixis* at which the embryos—somatic and parthenogenetic—are able to arise in seeds from haploid or diploid megaspores and their derivatives (Figure 25).

A good example of this state is the phenomenon of haploidy — the changing of the program of development of microspore cells and pollen grain from gametophytic to sporophytic one under the influence of definite environmental factors and the obtaining of new individuals without fertilization by means of different morphogenesis pathways: embryoidogenesis (direct or indirect through callusogenesis) and gemmorhizogenesis (homophasic reproduction—vegetative propagation) (Figure 33-36). This phenomenon was found in very different angiosperm species.

The data on gametophytic apomixis and androcliny, in our opinion, convincingly testifies to the preservation in evolution of morphogenetic potentials (reserves) of reproductive structures, realization of which, however, is displayed only under certain conditions. This is the reason for proposing that *in some species of flowering plants in nature the spores fail in their main functions connected with reproduction and dispersal*, as Pisyaukova assumed (1980). In this connection, it is necessary to remind once again all investigators who work in approaches and methods of generative and reproductive structure cultivation that in culture *in vitro* only those potencies and reserves that are typical of all organism cells in natural conditions are capable of being realized (Batygina, 1987a, 1994b, 2005a, b; Batygina et al., 1978).

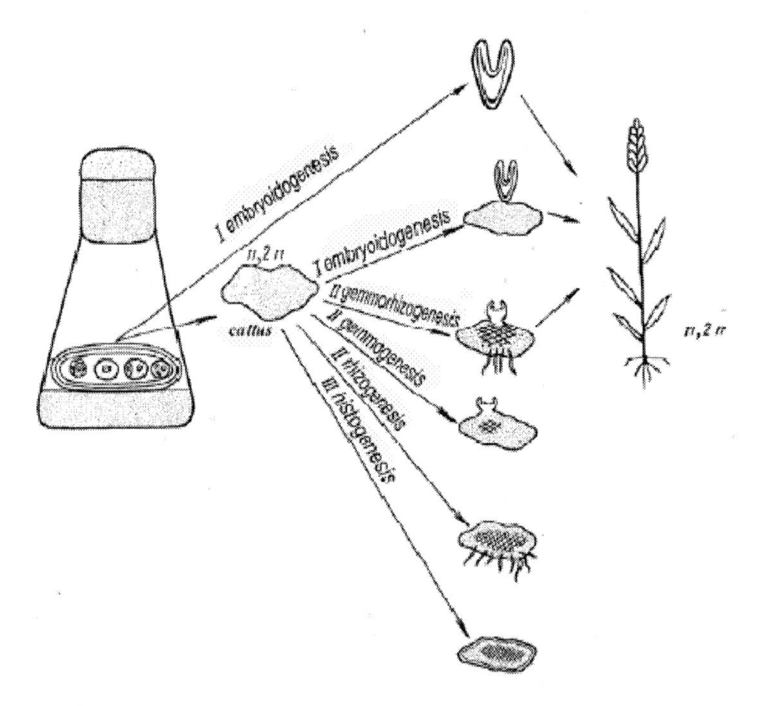

Figure 33. Possible pathways of morphogenesis in culture *in vitro* of anthers of flowering plants (from Batygina, 1984).

Vegetative embryoidogeny

Vegetative embryoidogeny is the formation of embryoids on vegetative organs; it includes foliar embryoidogeny (on leaf), cauligenous embryoidogeny (on stem) and rhizogenous one (on root). The arrangement of embryoid formation to the definite vegetative organ is taxon-specific.

Foliar Embryoidogeny

Two representatives of Crassulaceae family, *Bryophyllum pinnatum* and *B. daigremontianum*, were studied (Figure 37, 38). Uncompleted data concerning the morphogenesis of structures arising from vegetative organs results so that there is no united terminology for their identification. Thus, for example, the structures formed on the leaf of different *Bryophyllum* species are referred to as “buds” (Howe, 1931), “embryos” or “leaf embryos” (Naylor, 1932; Yarbrough, 1932; Warden, 1968), “leaf pseudobulblets” (Johnson, 1934; Resende, 1954), “embryoids” (Batygina, 1987a-c, 2006a).

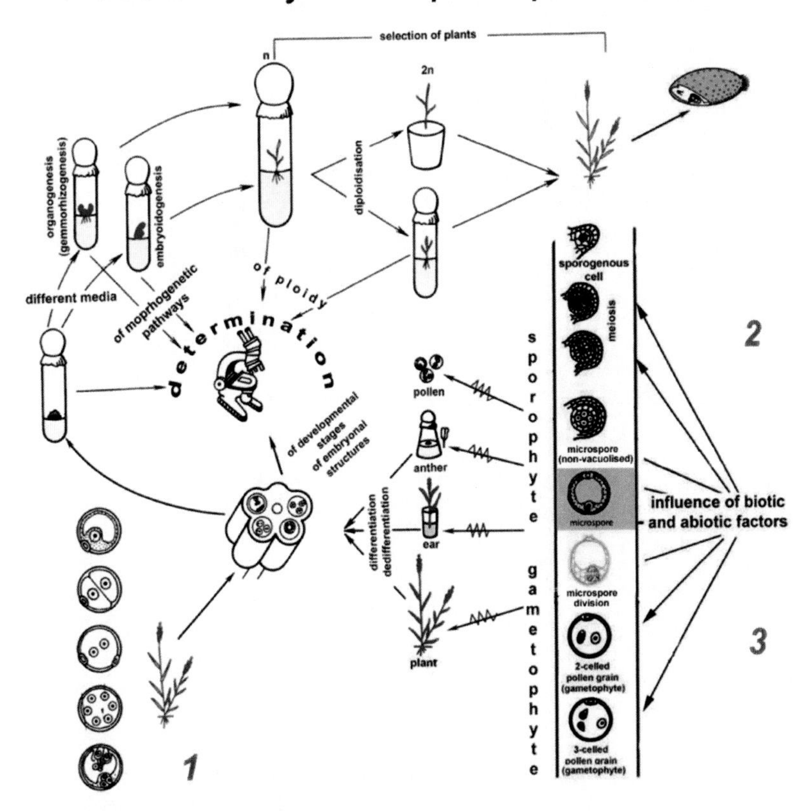

Figure 34. (1) natural abnormalities of microspores and pollen; (2) induced abnormalities; (3) impact on normal microspores, two- and three-celled pollen grains eliminating determination.

Producing of wheat haploid plants-regenerants by direct or indirect embryoidogenesis in vitro

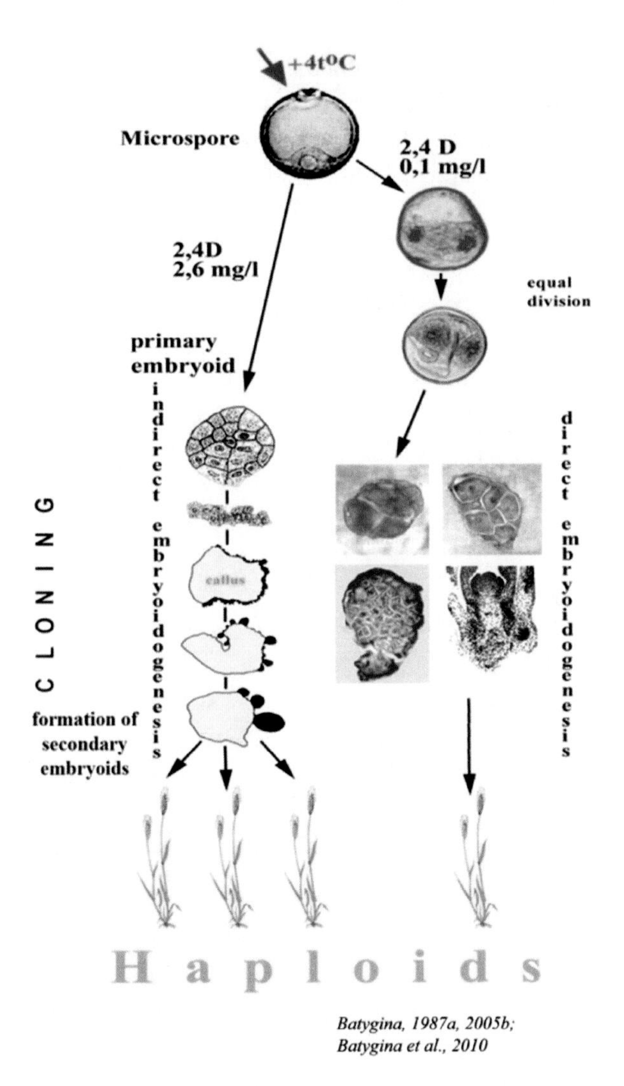

Figure 35. Perspective way of mass producing of wheat haploid plants-regenerants through direct and non-direct secondary embryoidogenesis (somatic embryogenesis) (from Batygina, 1987a, 2005b, 2010).

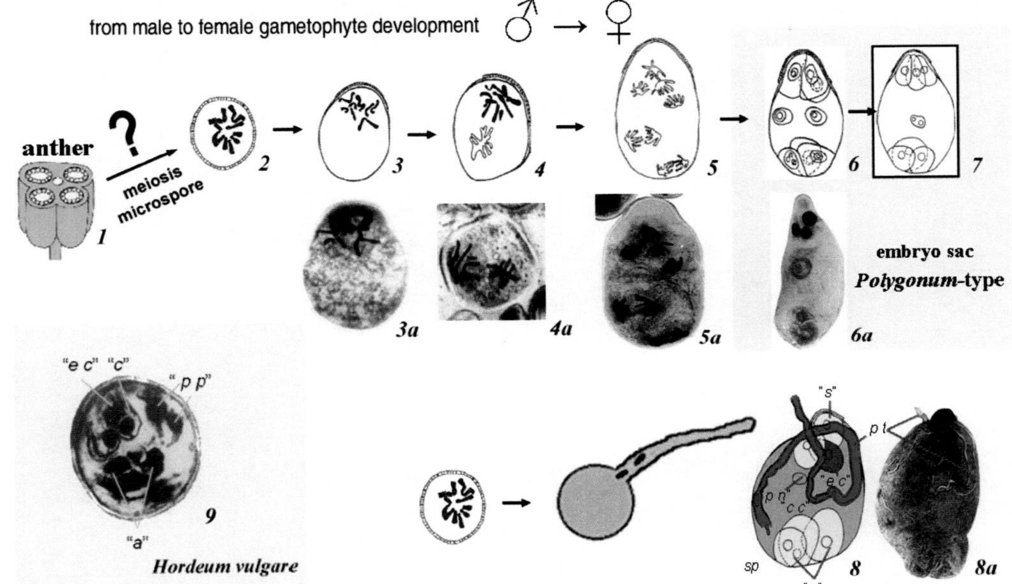

Figure 36. (1) anther; (2-7) consecutive stages of "embryo sac" of Polygonum-type development from microspore; (8, 8a) pollen tube entered "embryo sac"; (9) *Hordeum vulgare* "embryo sac" arisen from microspore. (2-7, 3a-6a – from: Stow, 1930; 8, 8a – from: Stow, 1934; 9 - from: Batygina, 1987).

a – antipodals, *c c* – central cell, *e c* – egg cell, *p n* – polar nuclei, *p t* – pollen tube, *s* – synergid, *sp* – sperm.

In one of the latest works concerning the research of reproduction in the genus *Bryophyllum* (=*Kalanchoë*), the authors (Garcês et al., 2007), while discussing the process of the structure formation on the leaf in different species, speak about *embryogenesis* which is, of course, not correct because this term is "occupied" for the development of sexual embryo to which the meiosis and sexual process precede. The embryoid on leaf derives from the leaf somatic cells. Therefore, they should be correctly referred to as embryoids, and the phenomenon of their formation is termed the phenomenon of embryoidogeny.

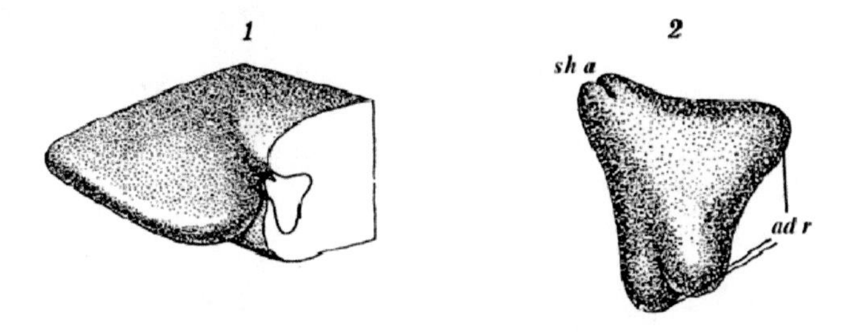

Figure 37. Leaf embryo *Bryophyllum pinnatum*: (1) embryoid inside the leaf tissue; (2) isolated embryoid with shoot apex (*sh a*) and adventive roots (*ad r*) (from Yarbrough, 1932).

The results of investigations of *Bryophyllum pinnatum* and *B. daigremontianum*, different in developmental biology, are the evidence in favour of the last term. In *B. daigremontianum*, the constant producing of propagules from the leaf stem cells is observed, and in *B. pinnatum* only in stress situations (Bragina et al., 1995; Batygina et al., 1996). In the species studied the leaf structures are formed endogenously, from the meristem, which Yarbrough (1932) called "dormant meristem". Temporal investigation we fulfilled differs from the works of other authors by that we observed for the first time the cells of "dormant meristem" (stem cells) when the leaf reaches only one millimetre. This zone in *B. pinnatum* is located *in leaf excision*, whereas in *B. daigremontianum*, *in excision projection* (Figure 38). It conditions the *endogenous formation of embryoids*. At early stage of leaf development the cells of meristematic zone do not contact with the conductive system. This is likely one of the reasons that it stays "dormant meristem". The contact between them *is established at the later stage of leaf development, after formation of 7–10 layers of cells with dense plasma in this zone.*

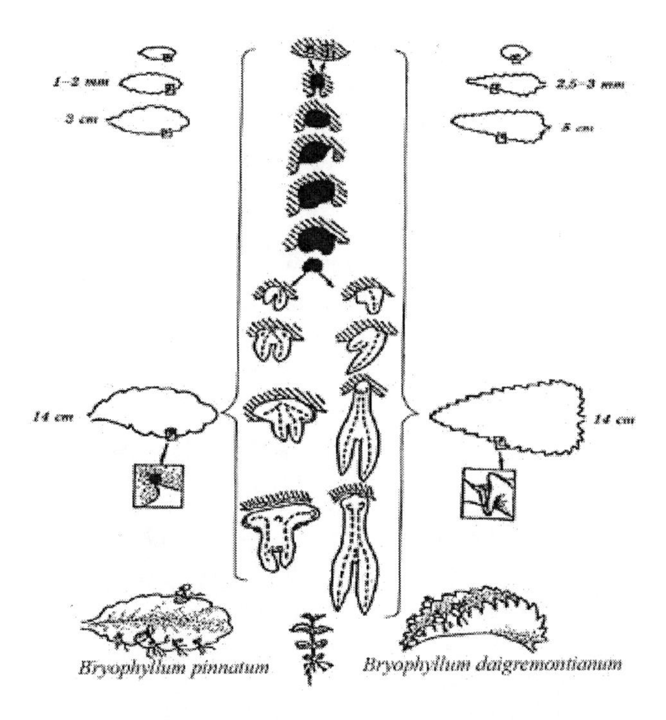

Figure 38. Development of the vegetative propagule on a leaf: (1) scheme; (2) the edge of a young leaf of 10 mm length with an indent on it, a marginal meristem and conducting bundle can be seen; (3) the edge of a mature leaf with an indent on it, meristematic zone, out of which the propagule and conducting bundle are formed; (4, 6) vegetative propagule at the heart-shaped stage of development; (5, 7, 8) vegetative propagule at the torpedo-shaped stage with two unequal cotyledons and two adventive roots (in the cavity of the section at [5] only one of them can be seen), the shoot apex has a poor morphologic manifestation; (7) view from above; (8) bottom view; (9) the edge of a young leaf of 15 mm length with the meristematic zone, out of which the propagule is formed; (10) vegetative propagule at globular stage of development; (11–13) vegetative propagule at the torpedo-shaped stage, two unequal cotyledons, two adventive roots and shoot apex can be seen; (14) prominence of the edge of a leaf, where the zone left after separating of the propagule can be seen; (2–8) *Bryophyllum pinnatum*; (9–14) *B. daigremontianum*; (2–5, 9–11) LM; (6–8, 12–14) SEM;

ad r – adventive root, *sh a* – shoot apex, *in* – indent, *m z* – meristematic zone, *c b* – conducting bundle, *ct* – cotyledon (from Batygina et al., 1996).

Later, the formation of new "leaf" structures (embryoids) begins. As the earliest stages of propagule development from dormant meristem are not identified (either they arise from one or several cells), one could speak about the resemblance of these structures with sexual embryo only at later developmental stages. The sexual embryo of dicots is known to undergo globular, heart-shaped, torpedo-shaped stages and the stage of maturation. *The same stages can be distinguished in the development of leaf structures in two species of Bryophyllum, but "the globular" stage differs by the morphology from classical globular stage of sexual embryo.* The formation of "cotyledons" in these structures occurs in common with sexual embryos of dicots. In both species one of "cotyledons" is usually smaller then another at early developmental stages. Later they became actually of identical sizes, the conductive system differentiates in them, and starch is accumulated. The plumule with two leaf primordia forms between cotyledons.

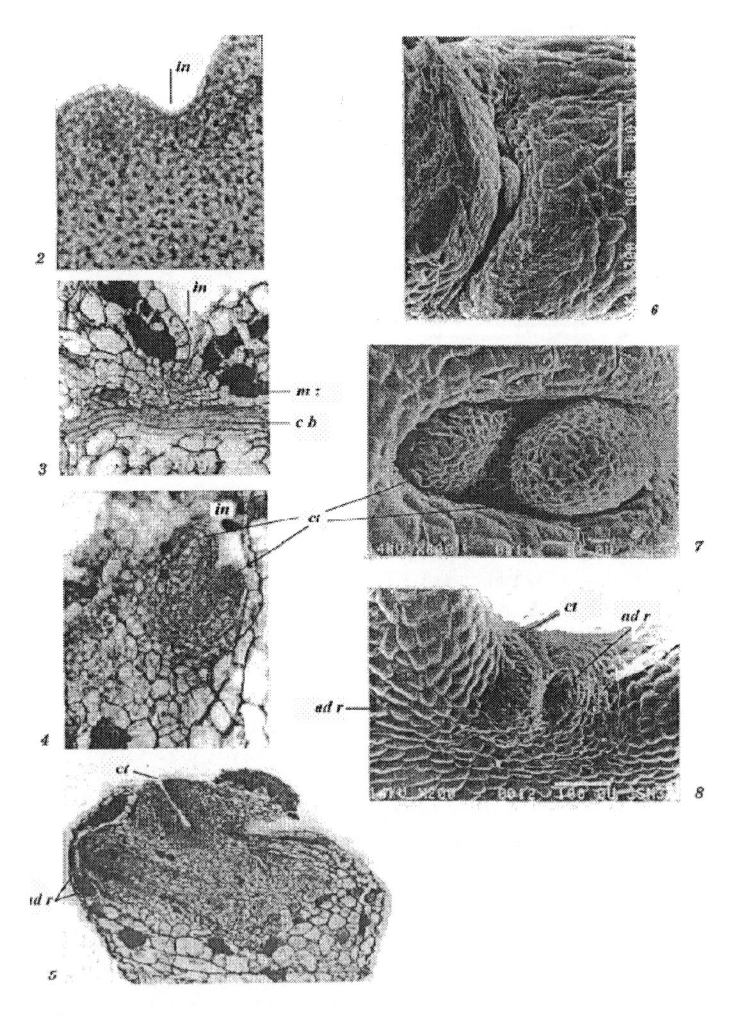

Figure 39. (Continued).

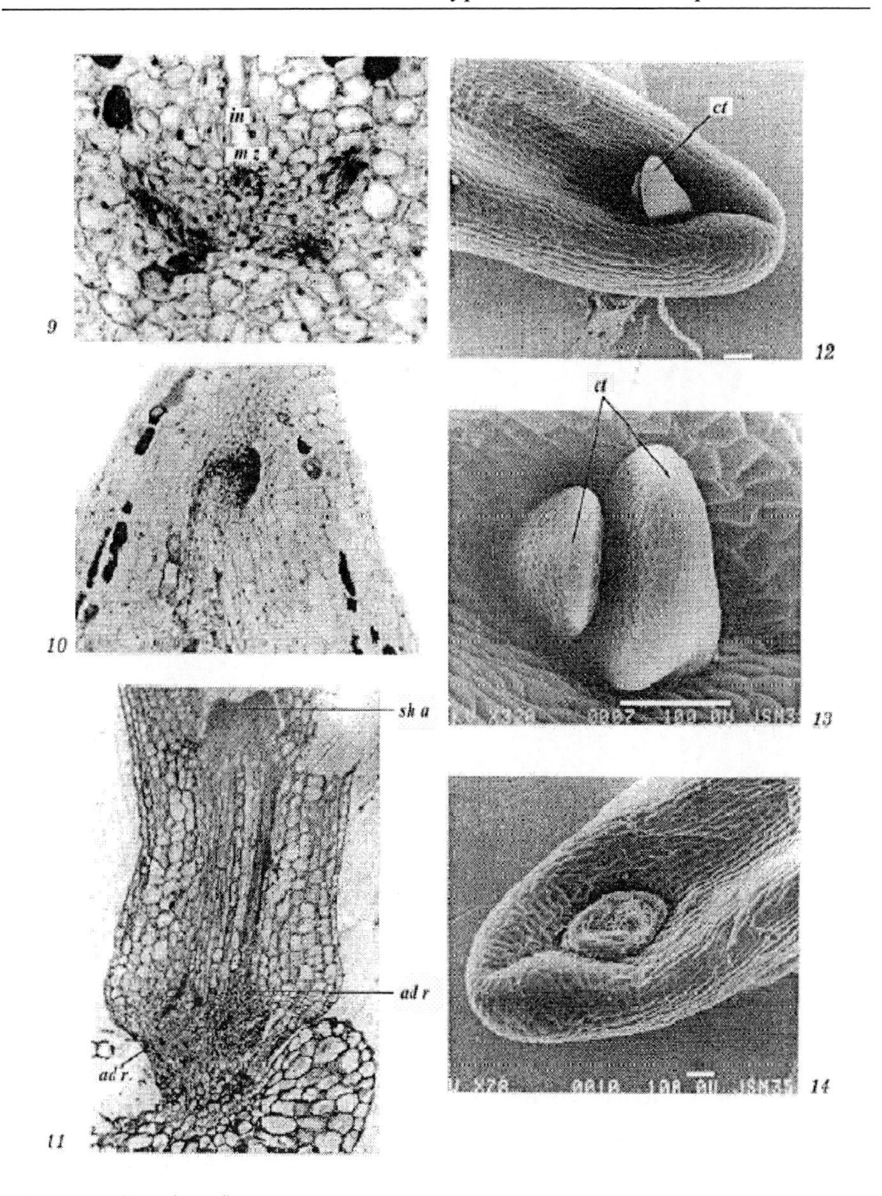

Figure 39. (Continued).

Figure 39. Development of embryo (embryoid) out of cells of the leaf epidermis in *Crassula multicava*. (1) appearance of the plant; (2) transversal section of the petiole; the divisions are noticeable in the epidermis cells 4 days after leaf separation; (3) longitudinal section of the petiole; the young embryo is noticeable; (4) longitudinal section of the leaf blade; the embryo is noticeable (from McVeigh, 1938).

The vegetative structures of *B. pinnatum* and *B. daigremontianum* are characterized by the lack of apical meristem of the main root, instead of which the system of adventive roots is formed. The laying down of adventive roots in *B. daigremontianum* is observed in the area of "hypocotyl" at the stage of late torpedo (well developed "cotyledons" with formed conductive system and the starch in their cells and the plumule with two leaf primordia). The laying down of adventive roots in the vegetative structure of *B. pinnatum* is observed at the stage when "cotyledons" are well developed, the conductive system begins to form in them, but the starch is even lack and the plumule is even not differentiated. Adventive roots are connected with leaf conductive system.

In the genesis of the vegetative structure in species investigated, the differences in the organization of its basal part, the time and the place of adventive root initiation were revealed.

Subsequently, the roots break through leaf tissues and come to opening excision. The separation of seedlings from the leaf happens due to the formation of separating layer of cells, the part of them is filled with tannins.

Vegetative structures can be found on the maternal leaf up to the stage when 6 pairs of leaves are formed in them. Vegetative structures of *B. daigremontianum* in condition of high humidity in long day are able to produce new vegetative structures on the third pair of leaves.

Comparable morphological and histogenic analysis of vegetative structures allows ones to consider these structures in the both *Bryophyllum* species to be similar. Such features as: the origin from dormant meristem; the proceeding of main developmental stages, inherent in sexual embryo (globular, heart-shaped, torpedo-shaped); the presence of cotyledons and well developed plumule; bipolarity of their development could be regarded as argument. The formation of adventive roots on the base of dormant meristem derivatives is the specifics of these foliar embryoids. Short-term connection of embryoid with the conductive system of leaf *does not mean yet, that this structure is a bud*, but indicates its *transition form (state) between embryoid and bud*. These data confirm our opinion (Batygina et al., 2006).

The diversity of modes of embryoid formation in different members of *Bryophyllum*, differed by constitution or induction of such formation, already speaks about the definite specifics of their reproduction biology and strategy of reproduction related with their ecology and centre of origin. The work of last times should be mentioned concerning molecular-genetic aspects of investigation of various *Bryophyllum* species, characterized by different peculiarities of their development. The genes *KdSTM, KdECI* and *KdFUS3* were separated for the first time in *B. daigremontianum*, the products of which in some species express at embryogenesis and organogenesis and by which in rather extent the blocking of the formation of seeds with normal developed embryo is determined (Garcês et al., 2007).

According to the new data of Garcês et al. (2007) on evolution of asexual reproduction in *Bryophyllum* genus, in the morphogenesis process in *B. daigremontianum* two programs work—organogenesis and embryogenesis (embryoidogenesis – T.B.). These authors consider that the induced formation of plants and viability of seeds in some species are the ancestral features; according to their opinion, the asexual reproduction is initiated as a process of leaf organogenesis and then as an embryogenesis (embryoidogenesis – T.B.) process in leaves in response to the loss of sexual reproduction in this genus.

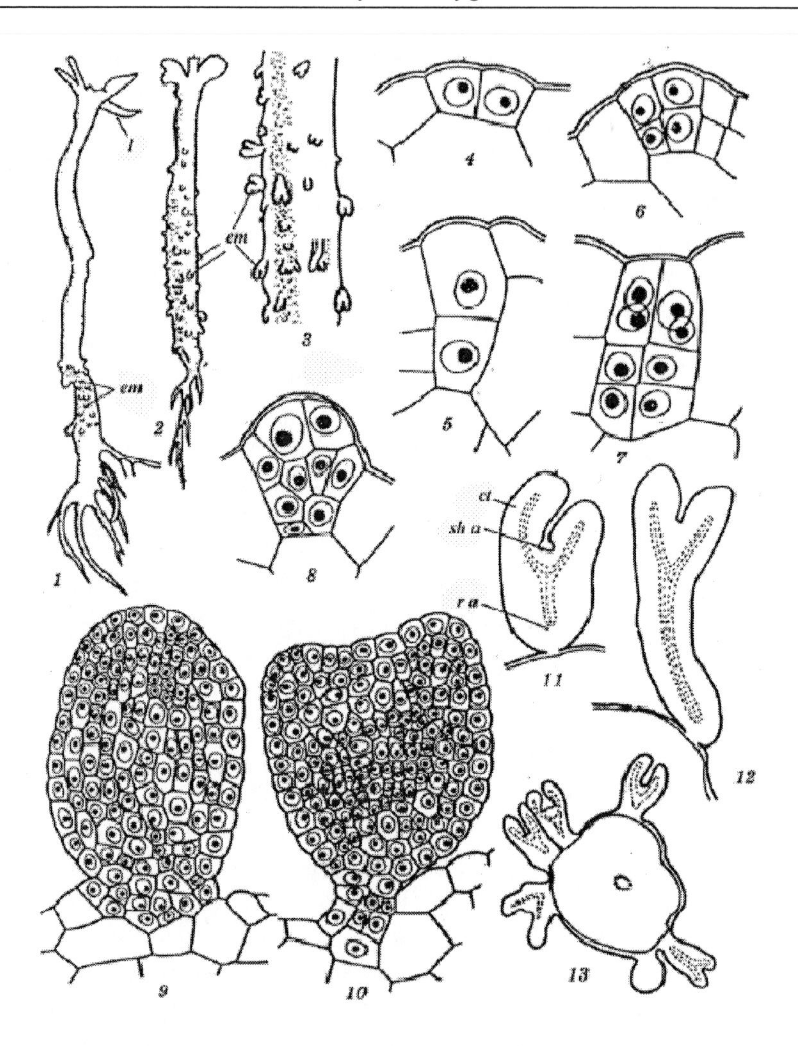

Figure 40. Development of somatic embryos (embryoid) out of cells of stem epidermis in *Ranunculus sceleratus*: (1, 2) one-month somatic embryos at lower (1) and upper (2) parts of the stem; (3) the part of the stem with embryoids; (4) two epidermis cells, which will produce embryoids; (5, 6, 7) two-, four-, and eight-celled embryoids; (8) more advanced stage of development of embryoid; (9) multi-celled embryoid; (10) somatic embryo with primordia of cotyledons; (11) somatic embryo with cotyledons and root apex; (12) the same, more advanced stage of development of embryoid; (13) transverse section of the stem with developing embryoids on it;

r a – root apex; *sh a* – shoot apex; *ct* – cotyledon; *em* – embryoids; *l* – leaves; (1–13) schemes; (from Konar, Nataraja, 1965).

This question is believed to be discussible and, in our opinion, the phenomenon of embryoidogeny (vegetative propagation) and formation of embryoids on leaves is the primary in evolution and the seeds, produced as a result of fertilization, the secondary one. It is also confirms by the evolution of asexual propagation in animal kingdom (Ivanova-Kazas, 1977). Because of this we could not agree with the view of Garcês et al. (2007) that the seed reproduction (double fertilization) is the primary (see also Takhtajan, 1964).

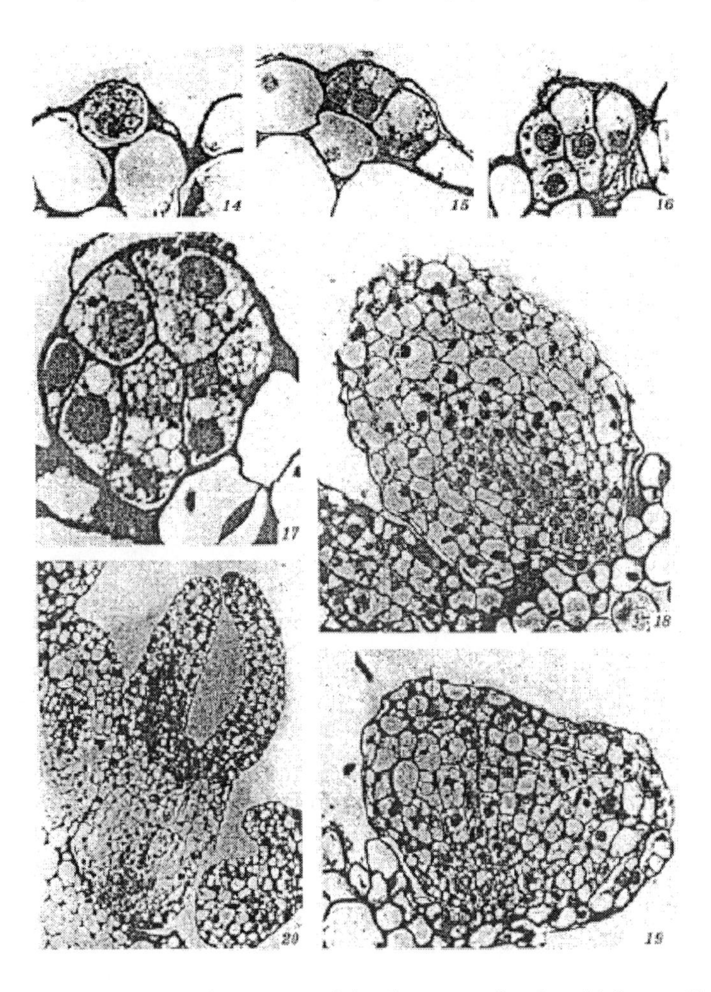

Figure 41. (14–20) Successive stages of development of embryoid from cells of stem epidermis in *Ranunculus sceleratus*; absence of its connection with the conductive system of the maternal organism, microphotos (from Konar, Nataraja, 1965).

In the other representative of Crassulaceae family — *Crassula multicava* (Figure 39) — the formation of embryo-like structures from the derivatives of epidermis cells in leaf or petiole (i.e., exogenously) was described (McVeigh, 1938). These bipolar structures as in the case of *B. daigremontianum* and *B. pinnatum* for a long time maintain contact with maternal organism.

Comparative analysis of the morphogenesis of several sexual embryos, propagules on leaves as well as embryoids formed in *in vitro* culture permits us to consider the using of notion "embryoid" for vegetative structures in different *Bryophyllum* species and in *C. multicava* to be more correct then using of other terms. Thus, we are of the same opinion as Naylor (1932), Yarbrough (1932) and McVeigh (1938), who suggested refer new vegetative structures formed on leaves to as "leaf embryos" or "embryos", but not as buds! *The term "embryoid" applied to Bryophyllum and in C. multicava we introduced in literature later.*

Most investigators compare leaf embryos with seeds (Yarbrough, 1934; Haccius, 1978). Dispersal of vegetative diaspores on different distances from maternal plant depends on the mode of their formation and the character of development. Seedlings arisen from these leaf embryos look like "tumbleweed".

Cauligenous embryoidogeny

Cauligenous embryoidogeny was firstly described in details in *Ranunculus sceleratus in situ* and *in vitro* (Konar, Nataraja, 1965; Konar et al., 1972). The embryoid of this plant as in *C. multicava* forms exogenously – from the derivatives of epidermis cells, its development proceeds similar to sexual embryo. At no stage of their development do cauligenous embryoids connect with the conductive system of the maternal organism (Figures 40 and 41), unlike the foliar embryos of *Bryophyllum* and *Crassula*.

Rhizogenous Embryoidogeny

In the literature, there are only fragmentary data, requiring additional investigations (see: Batygina, 1989a).

Parallelism in the Development of Sexual and Somatic Embryos

The resemblance and differences in the development of sexual and somatic embryos as well as the bud—elementary structural units of propagation — is the general problem of morphogenesis. For the first time the problem on the possibility of comparable study of sexual and somatic embryos obtained *in vivo* and *in vitro* and also the examination of these structures from gemmorhizogenesis positions was considered (Batygina, 1996b; Batygina, Zakharova, 1997a, b). In this relation, we attracted the materials from the fundamental investigations: *to what extent does the morphogenesis of the somatic embryo submit to the laws of development of sexual embryo—to the laws of origin, numbers, position and purpose and the law of economy as well as the laws of cell divisions* (Souèges, 1939; Johansen, 1950). Unfortunately, insufficient attention is paid to the role of somatic embryo (embryoid) as an elementary structural unit in plant reproduction system that is more economic and adaptive, especially for the problems of biotechnology and selection.

The terminology and classification of embryoids (adventive embryos) and "embryo-like structures" for a long time remains the subject of discussion.

The opinion arose that the embryoid lacks the main features typical of a sexual embryo, namely: the initial cell of somatic embryo has no polarity axis; the regular model of cell divisions is not observed in embryoid (irregular course of divisions); the laws of cell divisions are destroyed; lack of suspensor; the embryoderm initiation is destroyed; lack of initials and typical centres of polar organization—epiphysis and hypophysis; (often) lack of hypocotyle initials; the normal organization of main root is disturbed; lack of pleroma, periblema, etc. On the bases of these the conclusion was done that the embryoid is closer by the morphology to the adventive bud, but not to the sexual embryo.

However numerous literature and original data on the development of sexual and somatic embryos in natural conditions and in the culture *in vitro* testify that the features mentioned above not always could be used for the determination of morphological status both of the sexual embryo and embryoid.

The system approach to the investigation of reproductive systems allowed comparison of the formation types both of sexual and somatic embryos *in situ, in vivo* and *in vitro* and revealing the parallelism in their development. Let us examine from the positions of such an approach: initial cells of sexual and somatic embryos; sexual embryos of plants of different taxa *with various types of embryogenesis and seedling formation*; somatic embryos *differing on the place of arising* (seed, leaf, stem, root) and development on the plant *in natural conditions* as well as on the mode of the formation and development *in the culture in vitro.*

Initial Cells of Sexual and Somatic Embryos

The character of differentiation of initial cells in both sexual and somatic embryos to a great extent conditions the course of embryo- and embryoidogenesis. Identification of such cells continues to be one of the main problems of morphogenesis both of sexual and somatic embryos. In this relation, the solution to the following questions is considered immediate:

- What cell in the embryo sac could be considered an analog of somatic embryo's initial cell: egg cell (gamete), zygote (sporophyte) or perhaps another cells of gametophyte (synergids, antipodal cells—at gametophytic apomixis)?
- What period of the mitotic cycle do these cells undergo, and what is the temp of the cycle proceeding (in particular, interphase)?

The egg cell of some plant species in natural conditions is known to produce the parthenogenetic embryos—embryoids possessing their own morphogenetic algorithm. The questions regarding the developmental stages—differentiation, specialization, etc. in which the egg cell is able to produce the parthenogenetic embryo and what is the trigger—also remain debatable.

In this relation, the initial cells of somatic embryo *in vitro* are worthwhile compare not only with egg cell but also with zygote as well as with initial cells that in natural conditions produce somatic embryos in the seed (cleavage, nucellar, integumental), on the vegetative (foliar, cauligenous and rhizogenous) and generative (androclinous) organs.

Initial Cells of Sexual Embryo

In flowering plants of all investigated taxa, during embryo sac development according to Polygonum-type, the first stages of egg cell (gamete) formation are rather universal. The initials of egg cells are morphologically similar (size and shape can slightly vary) and totipotent (Figure 42). They have the nucleus centrally situated, typical vacuolization and a wall. In the course of development of *"the initial of egg cell"* into *"the formed"* and then in "the mature" egg cell, its sizes are increasing, the form is changing, and the polarity and symmetry are establishing. Thereby the property of totipotency is partially being lost. The "mature" egg cell is characterized by the specific vacuolization of cytoplasm, nucleus migration and wall modification in its apical part.

gametophyte → gamete → zygote → sporophyte

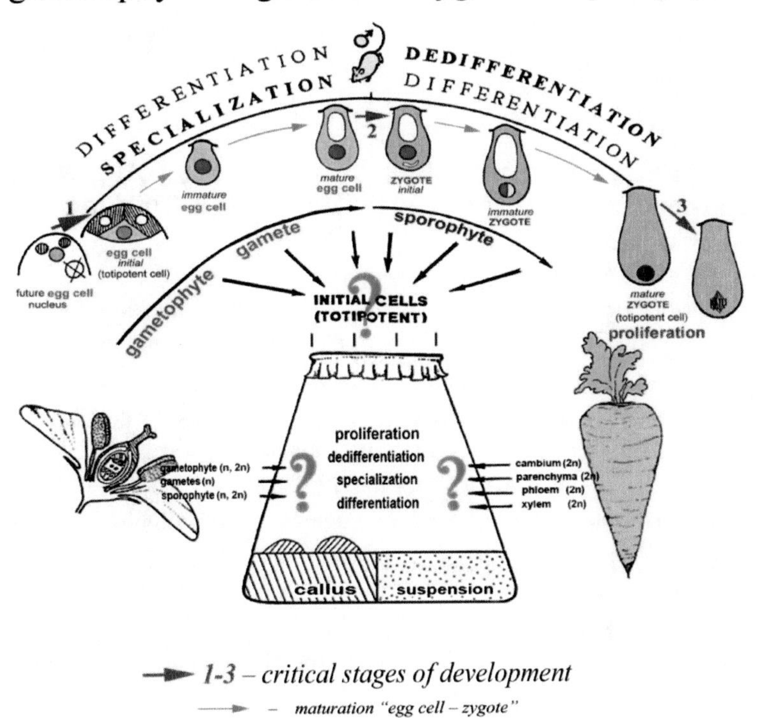

Figure 42. Multi-stage and pulsing character of egg cell transformation into zygote. Possible pathways of formation of initials cells of sexual and somatic embryos in natural conditions and in culture *in vitro*.

At the time of plasmogamy and karyogamy in the formed "zygote initial", the process of differentiation begins. It continues during zygote formation; its cytoplasm, wall and nucleus undergo a series of modifications resulting, on one hand, in the formation of a taxon-specific appearance of "mature zygote", and, on the other hand, in the obtaining again of meristematic state (totipotency) and capability to proliferation. So in the system "egg cell-zygote" the multistage processes occur: meristematization, differentiation, specialization, dedifferentiation and meristematization again (Figure 43). The egg cell and zygote are comparable with the initial of the somatic embryo only at the stages of its development characterized by its capability to division (proliferation). However, up to now it is not reliably established whether such processes happen in the cells producing somatic embryos *in situ, in vivo* and *in vitro*. It should be taken into consideration that environmental conditions have a great influence on the development of initial cells. The *initial of egg cell* as well as *immature* and *mature egg cell*, and also as zygote form in the complex system of embryo sac and gradually in the course of development become isolated and rather autonomous.

Initial Cells of Somatic Embryos (Monozygotic, Nucellar, Integumentary, Foliar, Cauligenous and Rhizogenous) in Natural Conditions

In various species of angiosperms, these cells are quite uniform by morphology and structure. They possess a series of common features with quickly dividing meristematic cells: little sizes, dense cytoplasm, large nucleus with noticeable enlarged nucleolus and many small vacuoles, and are characterized by high metabolic activity, intensive synthesis of RNA (Figure 44). The initial cells of somatic embryo before the first division usually obtain the polarity (the vacuole forms, and the nucleus is often situated eccentrically). In somatic embryos as in sexual ones initial cells are genetically determined, submit to the general laws of cell division and their further development occurs under the system of a whole individual. It is likely provides the normal formation of somatic embryo, seedling and the plant in natural conditions.

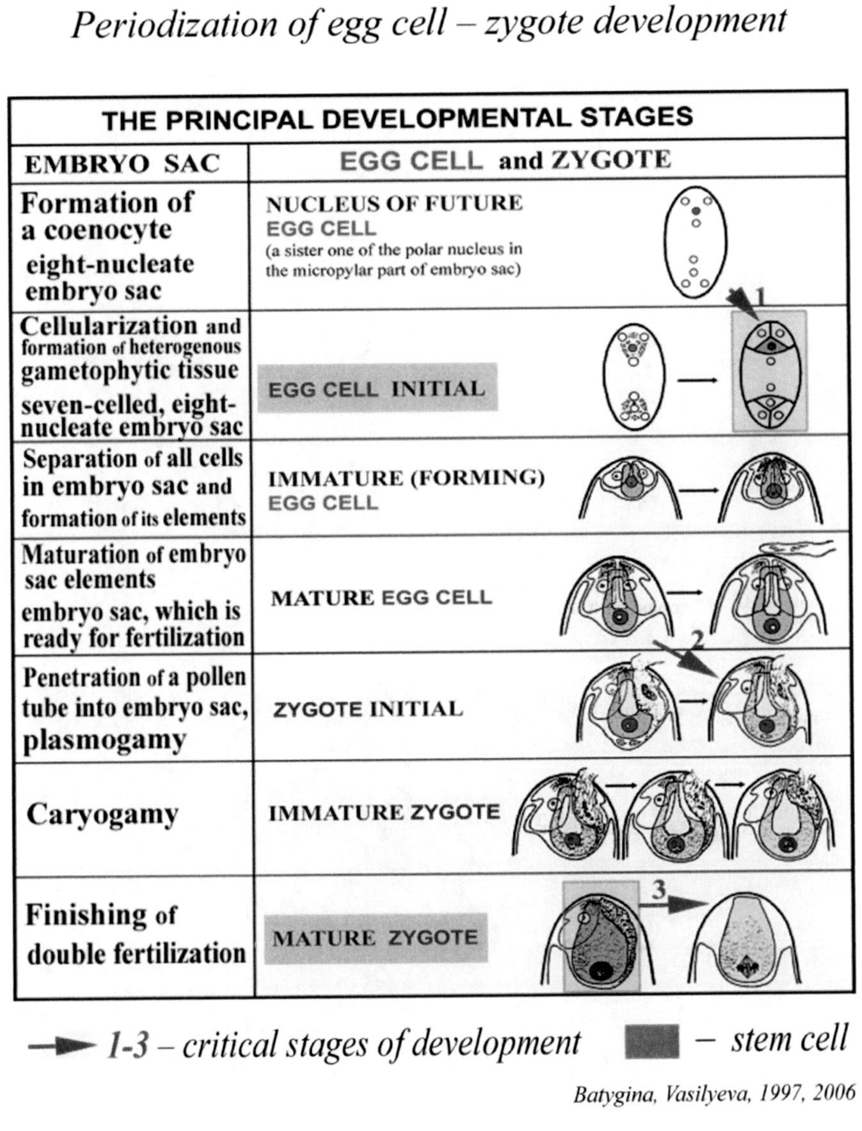

Periodization of egg cell – zygote development

THE PRINCIPAL DEVELOPMENTAL STAGES	
EMBRYO SAC	EGG CELL and ZYGOTE
Formation of a coenocyte eight-nucleate embryo sac	NUCLEUS OF FUTURE EGG CELL (a sister one of the polar nucleus in the micropylar part of embryo sac)
Cellularization and formation of heterogenous gametophytic tissue seven-celled, eight-nucleate embryo sac	EGG CELL INITIAL
Separation of all cells in embryo sac and formation of its elements	IMMATURE (FORMING) EGG CELL
Maturation of embryo sac elements embryo sac, which is ready for fertilization	MATURE EGG CELL
Penetration of a pollen tube into embryo sac, plasmogamy	ZYGOTE INITIAL
Caryogamy	IMMATURE ZYGOTE
Finishing of double fertilization	MATURE ZYGOTE

1-3 – critical stages of development — *stem cell*

Batygina, Vasilyeva, 1997, 2006

Figure 43. Periodization of egg cell – zygote development.

Initial Cells of Somatic Embryos in *In Vitro* Culture

Identification of such cells is one of the difficult problems of morphogenesis because, in most cases, the explant is heterogeneous in structure. The formation of somatic embryo in the culture *in vitro* is known to proceed by the pathway of direct or indirect embryogenesis. The direct embryogenesis is typical for the cells that have been determined even before their explantation; only trigger is required stimulating this process. Indirect embryogenesis occurs in the culture of cells capable to proliferation and callus formation.

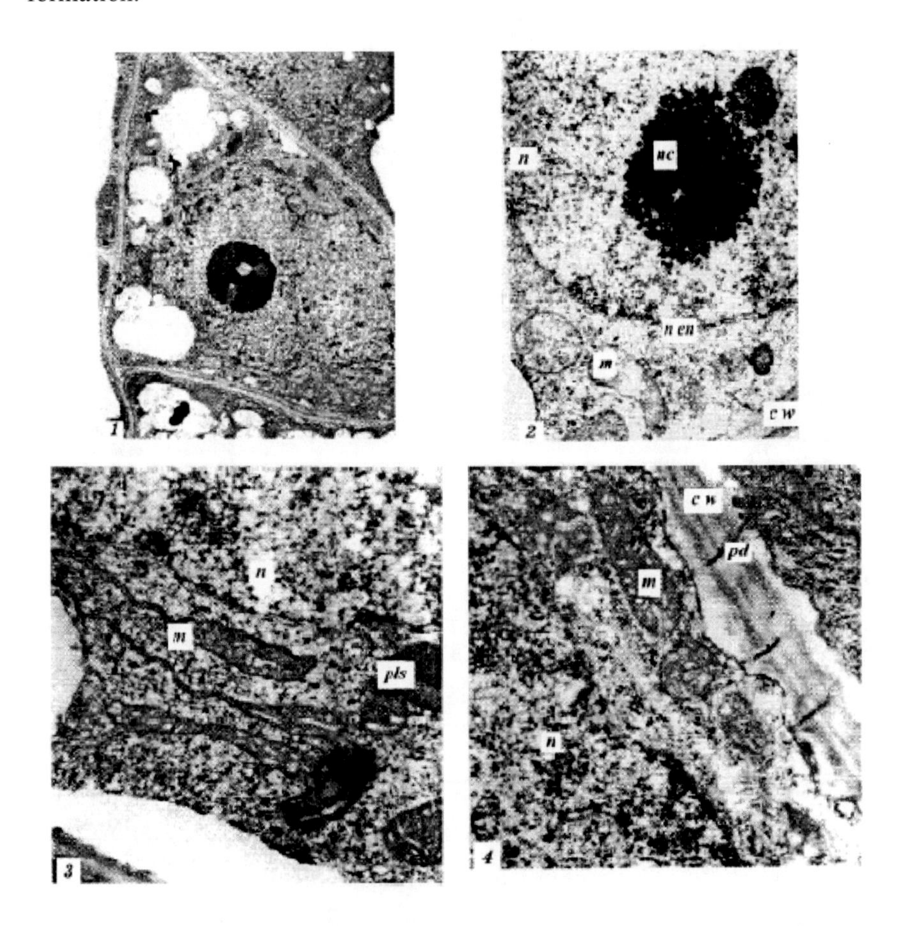

Figure 44. Initial cells of the embryoid in *Sarcococca humilis.* (1) embryocyte; (2–4) fragments of embryocytes: numerous pores can be seen in nuclear envelope (2), elongated profiles of mitochondria and oval ones of plastids are noticeable around the nucleus, polysomes can be seen (3), cell wall with plasmodesmata, mitochondria with numerous cristae (4);

c w – cell wall; *m* – mitochondria; *n en* – nuclear envelope; *pd* – plasmodesma; *pls* – plastid; *n* – nucleus; *nc* – nucleolus (from Naumova, 1993).

Critical mass of development

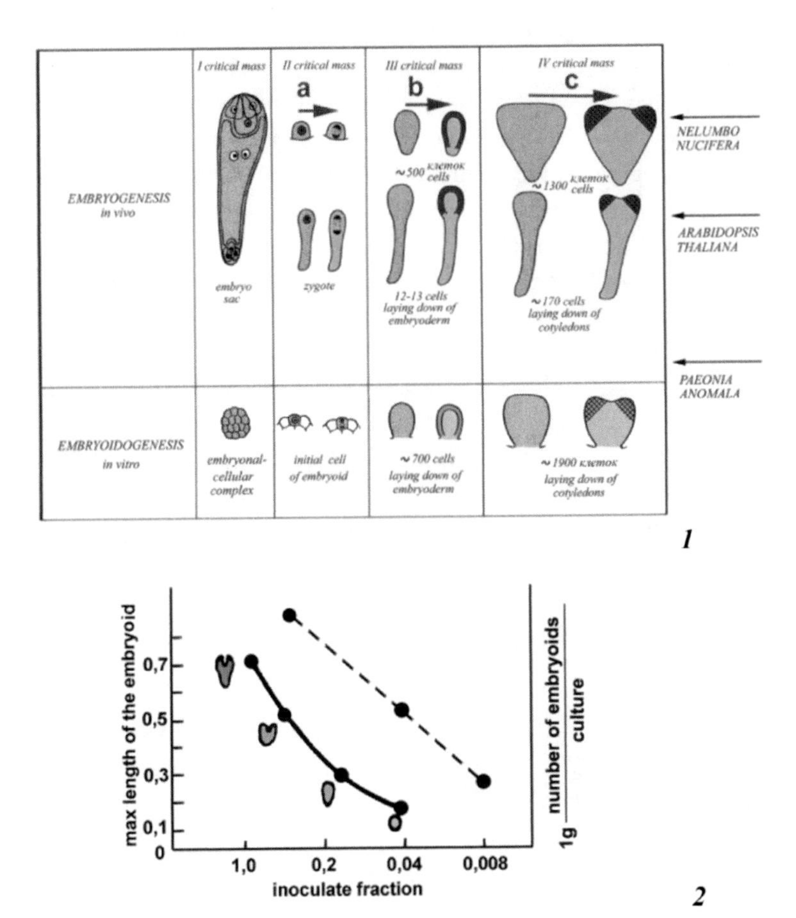

Figure 45. (1) Critical mass of cells during the development of sexual and somatic embryos (from Batygina, 1984); (2) Influence of density on differentiation of embryoids in *Daucus carota* (from Batygina, Vasilyeva, 2002).

There are two model systems, different in character of cell interaction, in tissue culture: callus and suspension cultures. In both heterogeneous model systems triggering of the processes of proliferation, differentiation and specialization of the initial cells and the embryo cells is to a great extent determined by the critical mass of cells and structures (Batygina, 1984) (Figure 45).

Callus is the heterogeneous structure (system), arising as a result of cell proliferation (mutations are possible) both on the wounded surface of certain structures of plant organism (exogenously) and in tissue depth (endogenously). More often it consists of heterogeneous cells (embryonic cell complexes, ECC), totipotency of which could be different, that conditions their various morphogenesis pathways (embryoidogenesis, organogenesis and histogenesis). The notion on callus as the amorphous structure or tissue arisen by irregular growth considers being not quite correct. The callus represents a heterogeneous integrated system (not obligatory tissue), because it derives as a rule from initially different cells of generative or vegetative organs (Figure 42). Its morphogenetic potencies are taxon-specific and can change in the process of genesis depending to rather extent on different factors (temperature, moisture, time of cultivation etc.). Some facts permit stating that morphogenetic potencies both of proper callus and its constituent elements—group of cells—and their regulatory capacity as well as morphogenesis pathways could be different: organogenesis, embryoidogenesis, gemmorhizogenesis or histogenesis (Figure 2). All this is to a great extent defined by the conditions of callus growing and also by the character of interactions between cell groups in callus, that, in its turn, is conditioned by their form, sizes (critical mass), ratio of different ECC, etc.

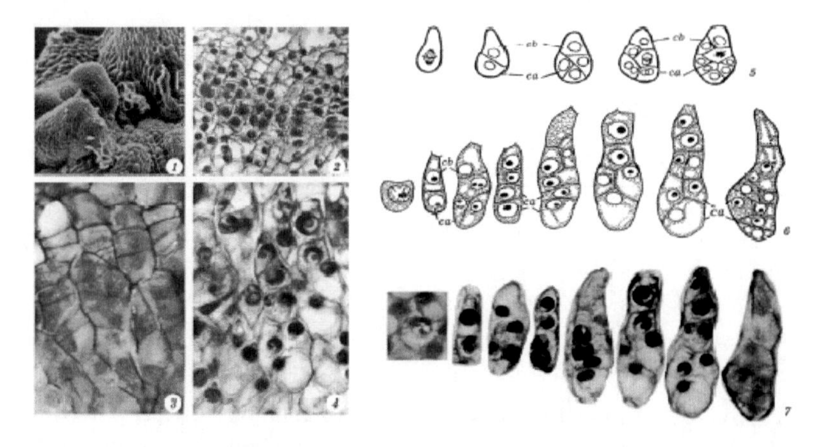

Figure 46. Endogenous formation of embryoids in *Triticum aestivum* in culture *in vitro*. (1, 2) callus tissue with meristematic zone, where the initiation of embryoids take place; (3, 4) initial stages of development of embryoids differently oriented in relation to the callus surface; (5–7) deviations in the development of somatic embryo (6, 7) in comparison with sexual embryo (5); (1) SEM; (2–4, 7) LM; (5, 6) scheme (Titova, Nikiforova, Batygina, orig. data).

It is likely that because of this there is no need to use so many various terms (*morphogenetic callus*, *centre of differentiation*, *zone of secondary differentiation*, *compact embryogenous callus*, *green callus*, etc.) for the identification cell groups in callus possessing regenerating potencies realized by different morphogenesis pathways: embryoidogenesis, gemmorhizogenesis, histogenesis etc. The notion "callus" is likely should be discussed, and one should call these structures *embryonal cell complex* as most scientists do (see Batygina, 1987a). Moreover, I consider to be expedient the addition to ECC the various definitions emphasizing its specifics, e.g., *ECC white* or *green*, *dense*, *compact*, etc.

In callus culture, the interaction between cells could be different. The polarity and division in these cells are usually defined by their location in relation to callus surface. So, for example, in callus culture of wheat (*Triticum aestivum*), the endogenous laying down of meristematic zone consisting from the layers of cells directly oriented in relation to callus surface was revealed. In the future, the cells of this zone become the initials of somatic embryos (Titova, Batygina, Nikiforova, orig. data, Figure 46). However, at cultivating of wheat microspores, exogenous embryoid formation was observed (initials, the cells of callus epidermis) (Figure 47, 34). In suspension culture (e.g., *Daucus carota*) the initial cells are polar before division, possess large vacuole and the nucleus driven to the cell wall.

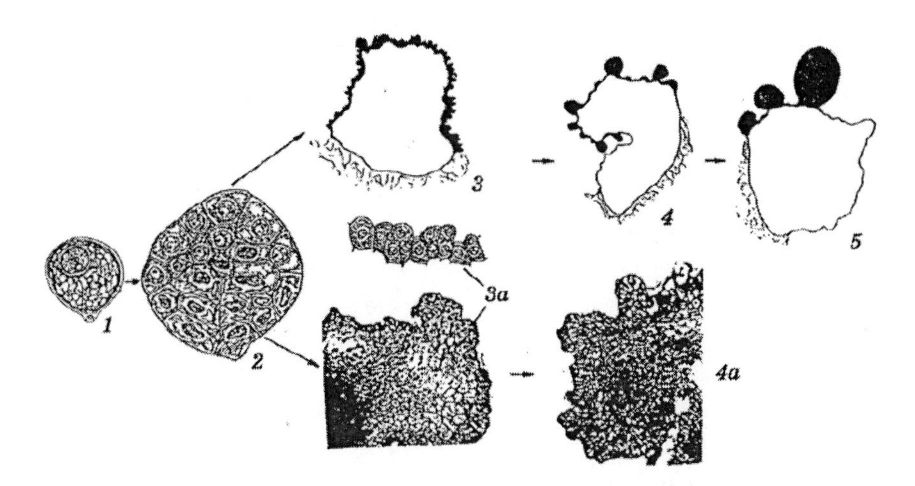

Figure 47. Exogenous laying down of initial cells of secondary embryoids in wheat callus. (1) microspore; (2) irregular primary embryoid; (3, 3a) callus, formed out of primary embryoid and exogenous laying down of initial cells of secondary embryoids; (4, 4a, 5) formation of secondary embryoids (from Shamrov et al., 1988).

The initial of the egg cell and the forming egg cell in the case of parthenogenesis or the mature totipotent zygote undergoing the complex pathway of morphogenetic modifications and capable of proliferation are likely to be considered the analog of the initial cell of diploid somatic embryo forming in culture *in vitro*.

At present, because of a lack of factual data on the morphogenetic processes in explant cells, it is very difficult to compare the development of the early somatic embryo in culture *in vitro* and egg cell–zygote in natural conditions. Nevertheless, on the basis of existing information, the definite parallelism in the development of initial cells of sexual and somatic embryos could be distinguished. The high degree of morphogenetic potency occurs to be the main property of all initial cells.

The obtaining of a *great number of somatic embryos and regenerants in different taxa in in vitro culture depends on the completeness of our knowledge of the genesis of initial cells* and their main characteristics (period of mitotic cycle, wall structure, state of cytoplasm, quantity and dispersal of organelles, etc.).

Sexual Embryo

Totipotency typical for mature zygote preserves in the cells *ca* and *cb*. While the embryo develops, in the process of histogenous differentiation and integration of certain cell complexes, it keeps only in the definite cell clusters (morphogenetic fields) of the embryo, seedling and plant. The significant diversity of *ca* and *cb* cells contribution in the embryo body construction should be mentioned. Using this feature, R. Souèges (1939) created the classification of embryogenesis types distinguishing six main types (megarchetypes). *He revealed in the embryo of some angiosperm species the epiphysis and hypophysis areas and their initials, the derivatives of which produce the apices of shoot and main root* (Figure 48). The formation of these regions as well as their initials usually takes place at early stages of embryo development, but the time of their differentiation is taxon-specific.

An onto-phylogenetic approach permits us to distinguish at least *five sexual embryo groups differing in this feature* (Batygina, 1997, 2006a, c):

1) the *typical initials of epiphysis and hypophysis* and *their centres* are present (e.g., *Geum rivale*);

2) only the *typical initial of hypophysis* and *its centre* are present (e.g., *Capsella bursa-pastoris, Arabidopsis thaliana*);
3) only *the typical initial of epiphysis* and *its centre* are present (e.g., *Polemonium caeruleum*);
4) the *typical initials of epiphysis* and *hypophysis* and *their centres are absent* (e.g., members of Poaceae family) (Figure 49-51);
5) *the presence of typical initials and centres of epiphysis and hypophysis in orchids* (e.g., *Gymnadenia conopsea*) *is debatable*.

In the embryos of the *first four groups, the morphological axis and polarity are formed, the shoot and root apices* (in the first three groups, the main root, and in the fourth, the adventive root) develop normally. In *the fifth group the mature embryo possesses polarity but the shoot and root apices are not expressed morphologically*.

Transitional Forms between Embryo and Bud

The analysis of literature and original data on the development of sexual and somatic embryos in the natural conditions has revealed that there are the structures differing from the typical sexual embryo and the bud by the morphology. Because of this we introduced the term “transitional form” under which we understand the structure possessing the features of embryo (e.g., the presence of globular, heart-shaped, torpedo-shaped developmental stages) and bud (the formation of adventive roots in the process of regeneration) (Batygina, 1993a, b). The embryos of such plants as *Nelumbo nucifera* (Nelumbonaceae), *Ceratophyllum demersum* (Ceratophyllaceae), the members of Poaceae, Orchidaceae, Orobanchaceae families illustrate the possible pathways of such transition.

In the embryos of *Nelumbo* and *Ceratophyllum* epiphysis and hypophysis are absent. In the embryo of *Nelumbo* (hydrophyte) while germinating the replacing of a main root by the system of adventive roots that initiate in the bases of bud leaves even at the late embryogenesis stages happens. But *Nelumbo* embryo from the very beginning of its existence occurs to be bipolar (Snegirevskaya, 1964; Kolesova, Batygina, 1988).

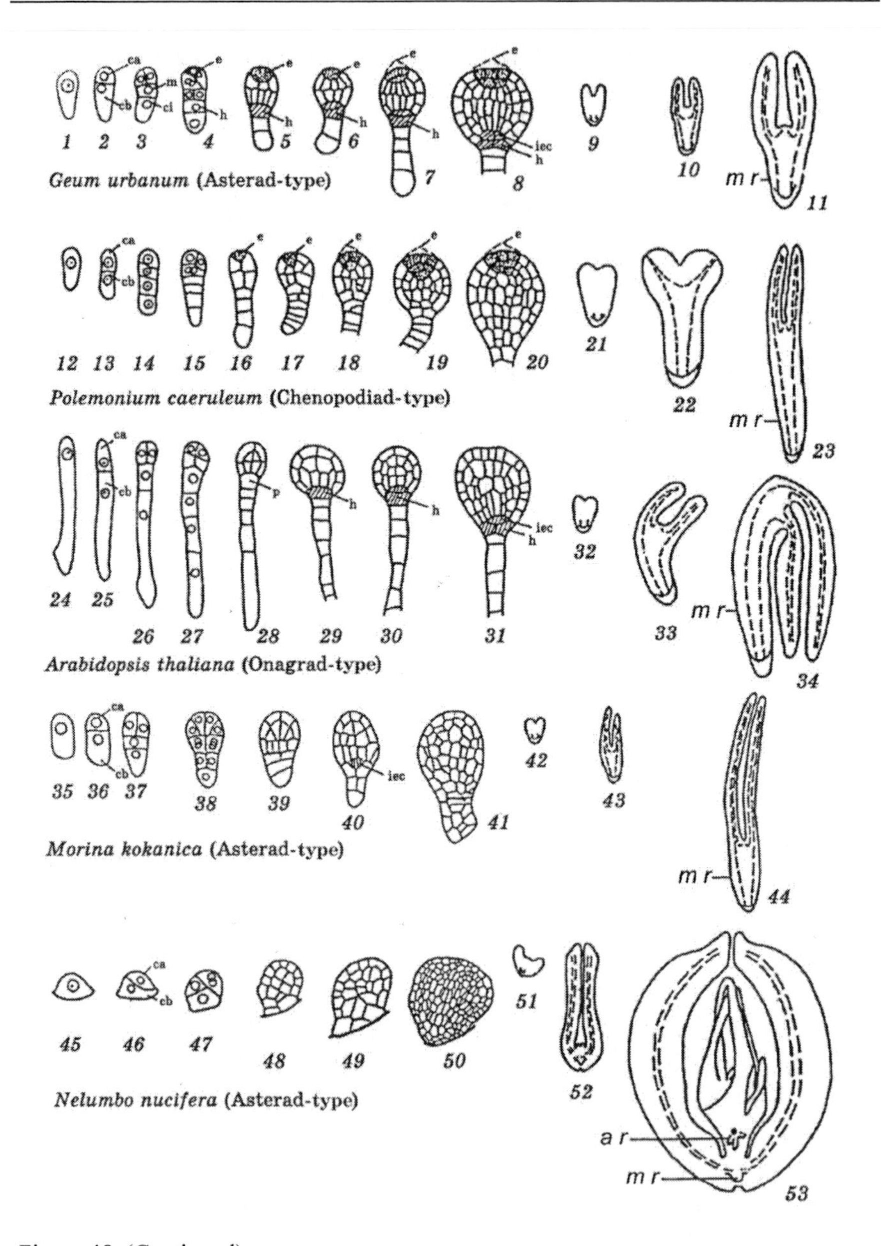

Figure 48. (Continued).

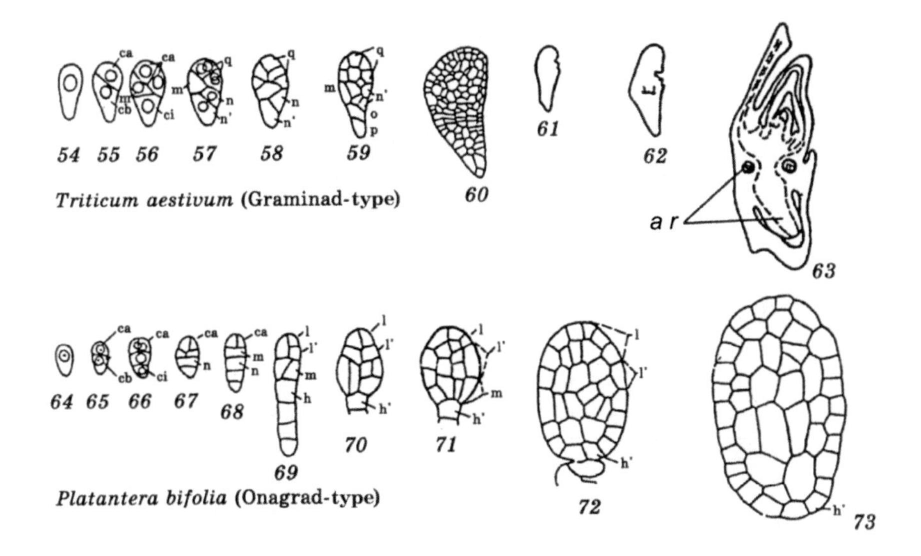

Figure 48. Polymorhism of sexual embryos in natural conditions: (1–73) successive stages of development of embryos with different degree of differentiation of epiphysis and hypophysis in dicotyledonous and monocotyledonous plants; (11, 23, 34, 44, 53, 63, 73) embryo in mature seed;

points show epiphysis, epiphysial centre (e); shading—hypophysis, hypophysial centre (h) ; *a r* – adventive root, *m r* – main root (from Batygina, Zakharova, 1997a).

The embryo of *Ceratophyllum* (hydatophyte) is likely also bipolar at the first stages of its development owing to the group of cells which could be considered as the initials of the main root. In mature embryo the only well developed bud is present and the main root is absent, at the same time the adventive roots do not form both in seminal and postseminal development. In the seedling of *Ceratophyllum* the typical bipolarity is absent.

In most representatives of Poaceae family (xerophytes) the formed bud and well developed adventive roots (their number varies depending on plant species, place and time of growing, e.g., frost and spring) are present in the mature embryo. Some authors assume the main root exists whereas another ones suppose, that in evolution process it had been modified in coleorhiza (Figures 50, 51) (see Batygina, 1974, 1987a).

In members of Orchidaceae family the mature embryo lacks the typical shoot and root apices although it is bipolar (difference of two zones). Later in the process of seedling (protocorm) formation together with the bud the adventive root appears. As a result the secondary polarity establishes (Teryokhin, 1977; Batygina, Vasilyeva, 1983; Batygina, 1993).

STRUCTURE OF EARLY WHEAT EMBRYO IN ***DIFFERENT PROJECTIONS***

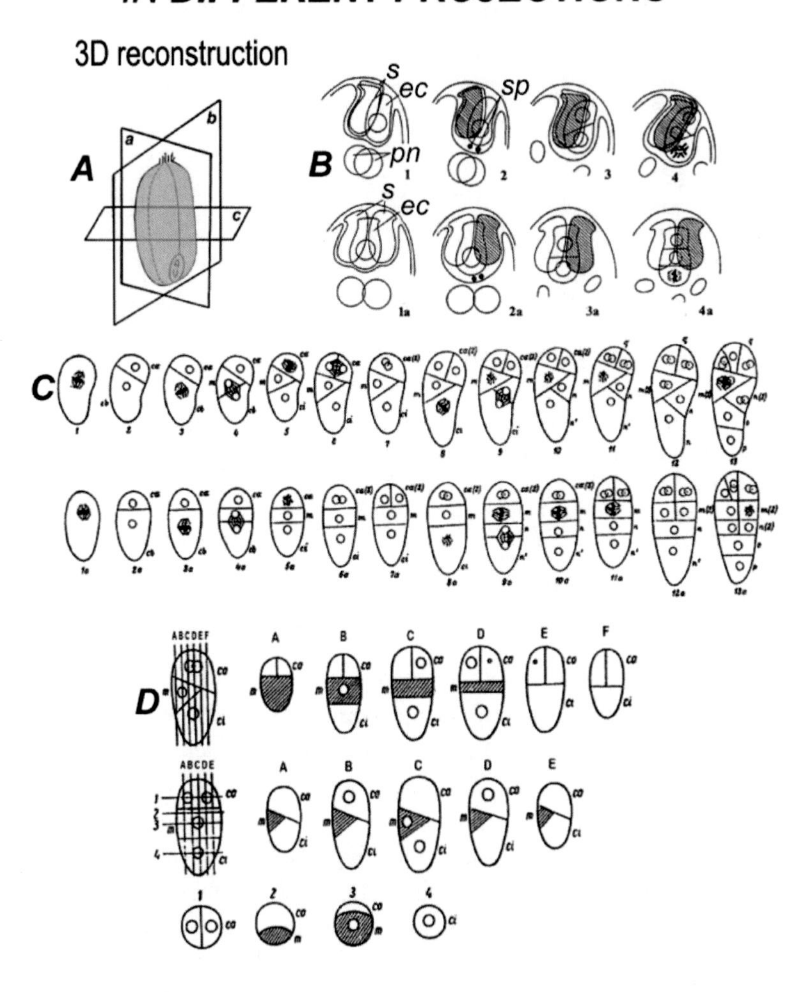

Figure 49. A – different sections of wheat kernel (*a* – dorsoventral, *b* – lateral, *c* – transversal); B – micropylar part of embryo sac in dorsoventral (1-4) and lateral (1a-4a) projections, before fertilization (1, 1a), at the moment of fertilization (2, 2a), at 2-celled (3, 3a), 3-celled (4, 4a) and 4-celled embryo stages; C – subsequent developmental stages of early embryo in dorsoventral (1-13) and lateral (1a-13a) projections; D – structure of four-celled embryo in lateral, dorsoventral and transversal projections.

Morphogenesis of Triticum embryo

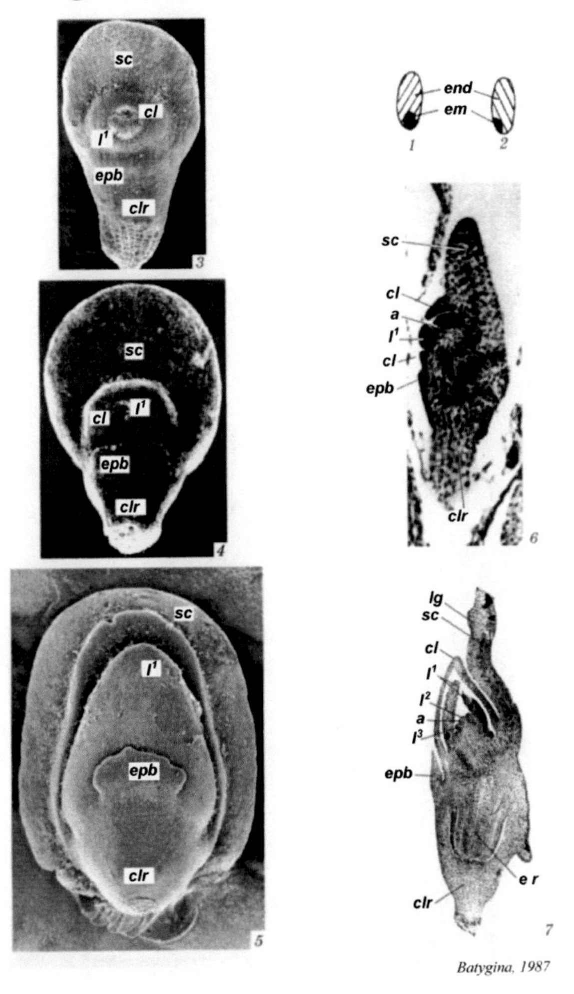

Figure 50. (1, 2) position of embryo in kernel (1 – full face, 2 – side view); (3, 4) general view of embryo taken from immature kernel; (5) general view of embryo taken from mature corn; (6) longitudinal section of embryo from immature kernel; (7) longitudinal section of embryo from mature kernel

a – apex, *cl* – coleoptyle, *clr* – coleorhiza, *em* – embryo, *end* – endosperm, *e r* – embryonic root, l^1, l^2, l^3 – first leaves of the plumule, *lg* – ligula, *sc* – scutellum.

Embryo development in Triticum sp.

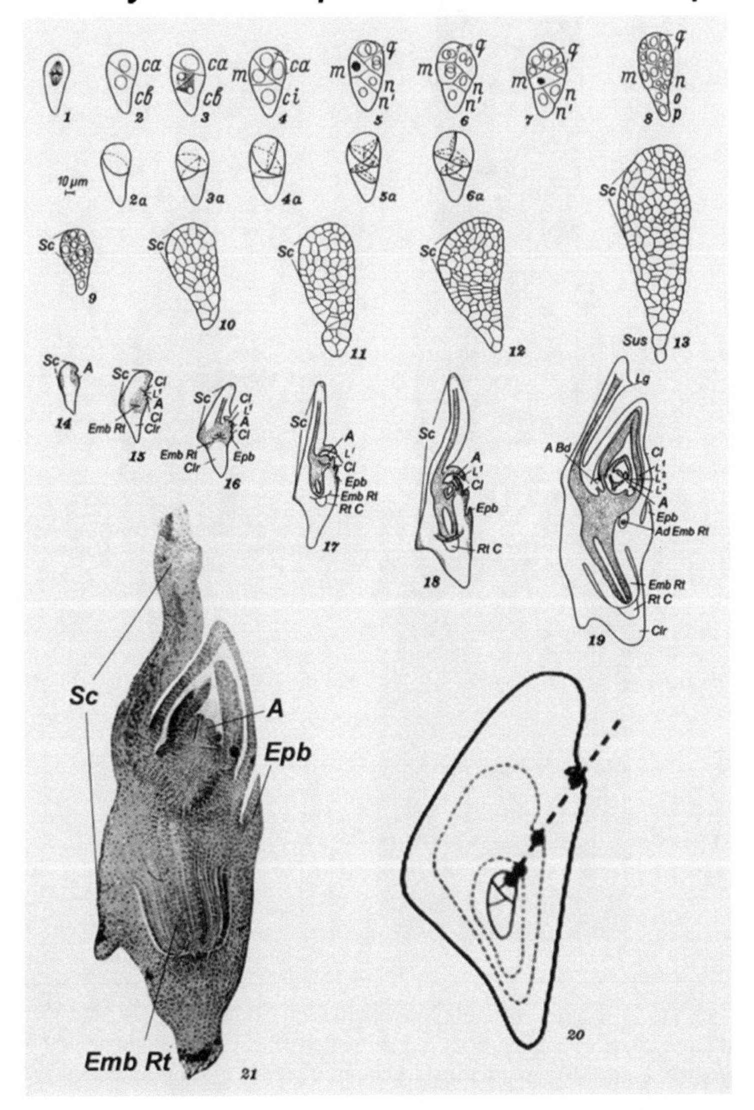

Figure 51. Points (14-19) show zones of mitotic activity in *Triticum*, 3d reconstruction.

In parasitic plants (Orobanchaceae, etc.) the bipolarity of embryo with different reducing degree may be observed from the very early stages of its development but later it loses. The secondary polarity appears only at seed germination (in contact with plant-host) thereby the true roots do not produce.

The embryo of parasitic plants represents the fine model for the studying of the reduction of *the typical initials of epiphysis, hypophysis as well as shoot and root apices* (Teryokhin, 1977). The great variability of the first developmental stages which displays in the different degree of *ca* and *cb* cell derivative participation in embryo body construction believes to be the peculiarity of its development. Often the first divisions are irregular but the development of this embryo does not allow applying it to any type of embryogenesis.

The series of sexual embryos of flowering plants given shows that the morphogenesis of sexual embryo is conditioned by the origin of its initials (zygote or egg cell in the case of parthenogenesis), type of embryo sac, ovule type and also by the peculiarities of taxon developmental biology. *The plasticity of embryo cells is rather high that is likely conditioned by the high degree of their morphogenetic potency.* This property makes possible *the deep morphogenetic modifications at different stages of embryo and seedling development with the preserving of their general organization.* These modifications in the process of adaptive evolution took place perhaps as a result of the disturbance of definite morphogenetic correlations at various hierarchy levels. The consequences of these modifications could be manifested at the very first (irregularity of cell divisions, different participation of *ca* and *cb* derivatives in embryo body construction etc.), medial (lack of hypophysis and epiphysis) and late (lack of the main root and adventive roots not only in embryo but also in seedling) stages of embryo development as well as in the general reducing of embryo differentiation level, beginning with early stages of its development. In the process of sexual embryo development the time of the appearance of primary polarity changes and also the replacing of it by the secondary polarity in course of seedling formation becomes possible.

It is possible to interpret the entire complex of features listed above as evidence of potential evolutionary trends towards the transition from classical sexual embryo to the bud or from the bud to the sexual embryo (Batygina, 1990a, b, 1993a, b).

Embryoid

Morphogenesis of somatic embryo (occasionally and a seedling) in natural conditions occurs in the system of the maternal organism, under its influence. Perhaps it conditions the normal proceeding of the whole embryogenesis and a high percentage of seedling yield, unlike the system "the culture of somatic embryo *in vitro*" where the percentage of normal plant-regenerants is not high.

The Development of Embryoid in Natural Conditions (Natural Embryoidogeny)

The embryoid of *Paeonia* members formed as the result of cloning of epidermal stem cells of sexual embryo occurs to be the bright example of natural embryoidogeny (monozygotic cleavage) (Figures 26, 27, 28, 52). Peony somatic embryo by morphogenesis character is similar to the typical sexual embryos of dicots – parallelism in development. The first stages of embryoid development proceed analogously to Asterad-type of embryogenesis (Figure 52, [1-21]). At initial cell stage the establishment of polar axis begins. Later the normal development of all histogens of root and shoot apices happens. The organization of somatic embryo in peony seed resembles the classical model of sexual embryo. In somatic embryos at the nucellular and integumentary embryoidogeny the initials of epiphysis and hypophysis and their centres are usually absent but the development of endosperm, shoot and root apices, the formation of a main root with all its elements proceed normally as in sexual embryos do, however the question on the character of the first embryo divisions stays debatable (regular or irregular).

Another mode of embryoid formation in Crassula multicava, where it derives from epidermal leaf cells, is considered interesting. In this case the typical initials of epiphysis and hypophysis and their centres are not revealed, the development of shoot and root apices in these embryos proceeds similarly with sexual ones (Figure 39).

In different *Bryophyllum* species the structures formed on the leaf have no initials of epiphysis and hypophysis and their centres. But during their development, as was noted above, "globular", "heart-shaped" and "torpedo-shaped" stages could be observed that speaks most probably about their belonging to somatic embryos or to the transitional form between embryo and bud. Morphogenesis of these structures is in rather extent similar to that of sexual embryo in grains (formation of shoot apex and adventive roots) (Figures 52, [22-29]).

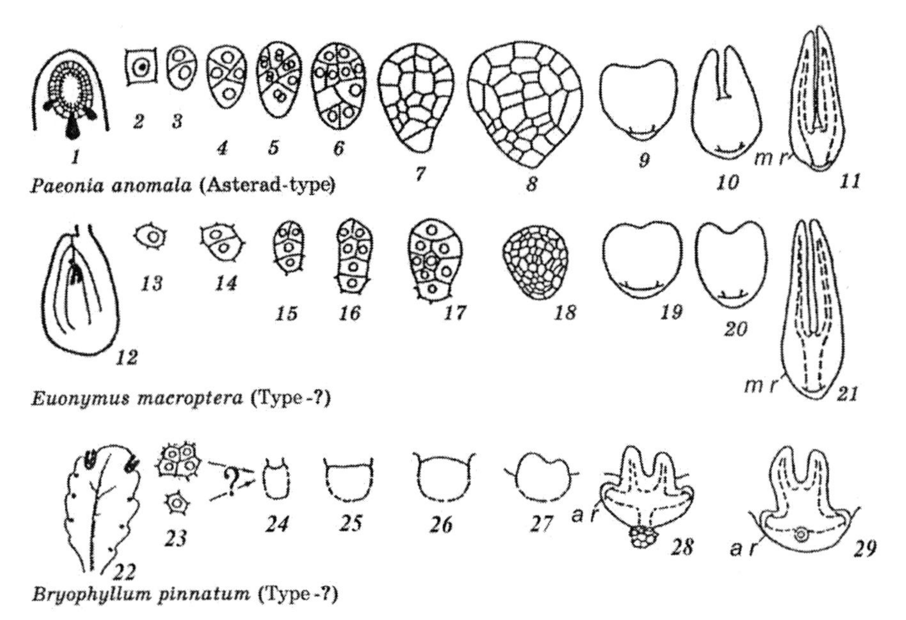

Figure 52. Polymorphism of somatic embryos in natural conditions: (1–29) successive stages of development of embryoids (1–11, cleavage; 12–21, integumental; 22–29, foliar); *ad r* – adventive root, *m r* – main root (from Batygina, Zakharova, 1997a with alterations).

Thus, the structure and genesis of embryoids are taxon-specific and are likely in a great extent conditioned by the place of initial cell formation and that conditions at which the development of individual and all population in a whole occurs. Morphogenetic resemblance of some embryoid types with the adventive bud confirms the existence of transitional forms between these structures.

The Development of Embryoid in *In Vitro* Culture (Artificial or Experimental Embryoidogeny)

As it was mentioned, *in vitro* culture presents to the investigators two different model systems for somatic embryo obtaining—suspension and callus. Let us examine the morphogenesis of such embryos on the example of different species of flowering plants contrasted by series of features. At the endogenous laying down of embryoids in the callus of *Triticum aestivum* the first divisions proceed similarly as in the sexual ones – Graminad-type of embryogenesis. However, the disturbance of the first pattern of cell division of their initial cell (equal, symmetrical instead of unequal, asymmetrical) and accordant further lying down of cell walls were observed more often that leads to abnormal histogenesis and the death of embryoids formed (Figure 46). The same picture of developmental disturbances is seen in the sexual plants of grains at distant hybridization.

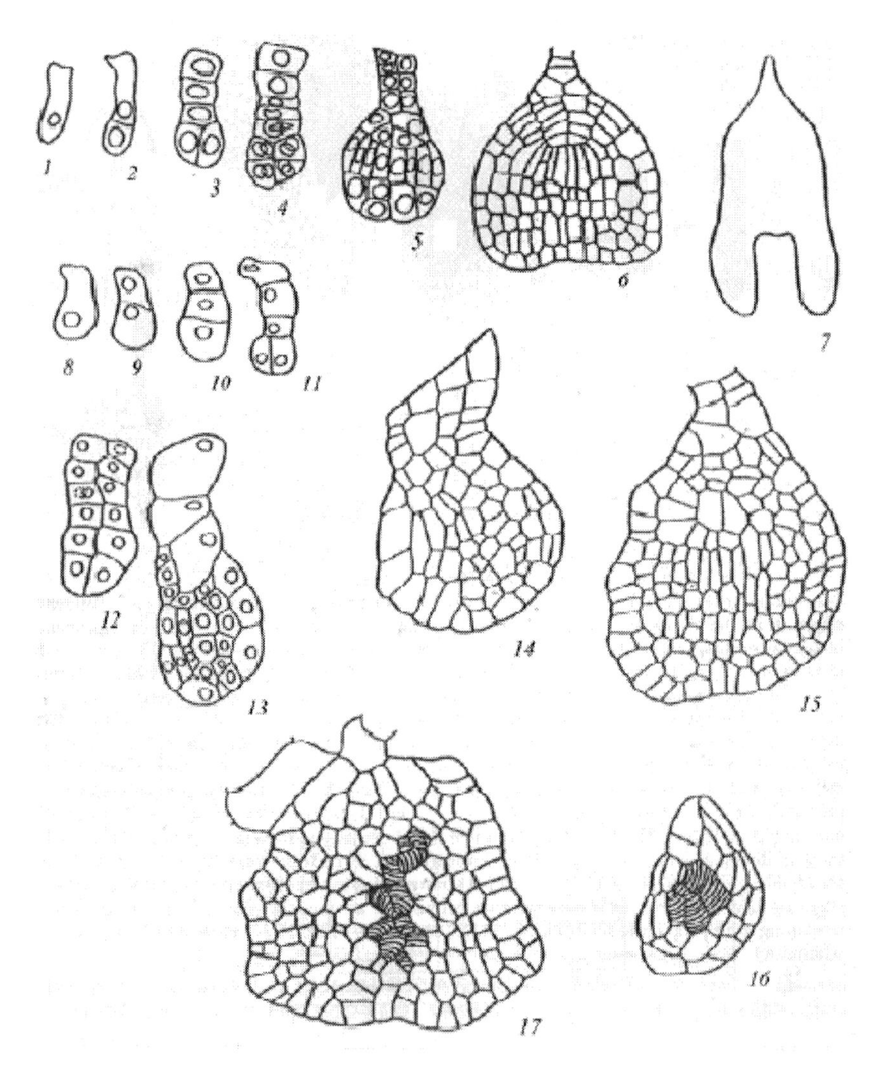

Figure 53. (Continued).

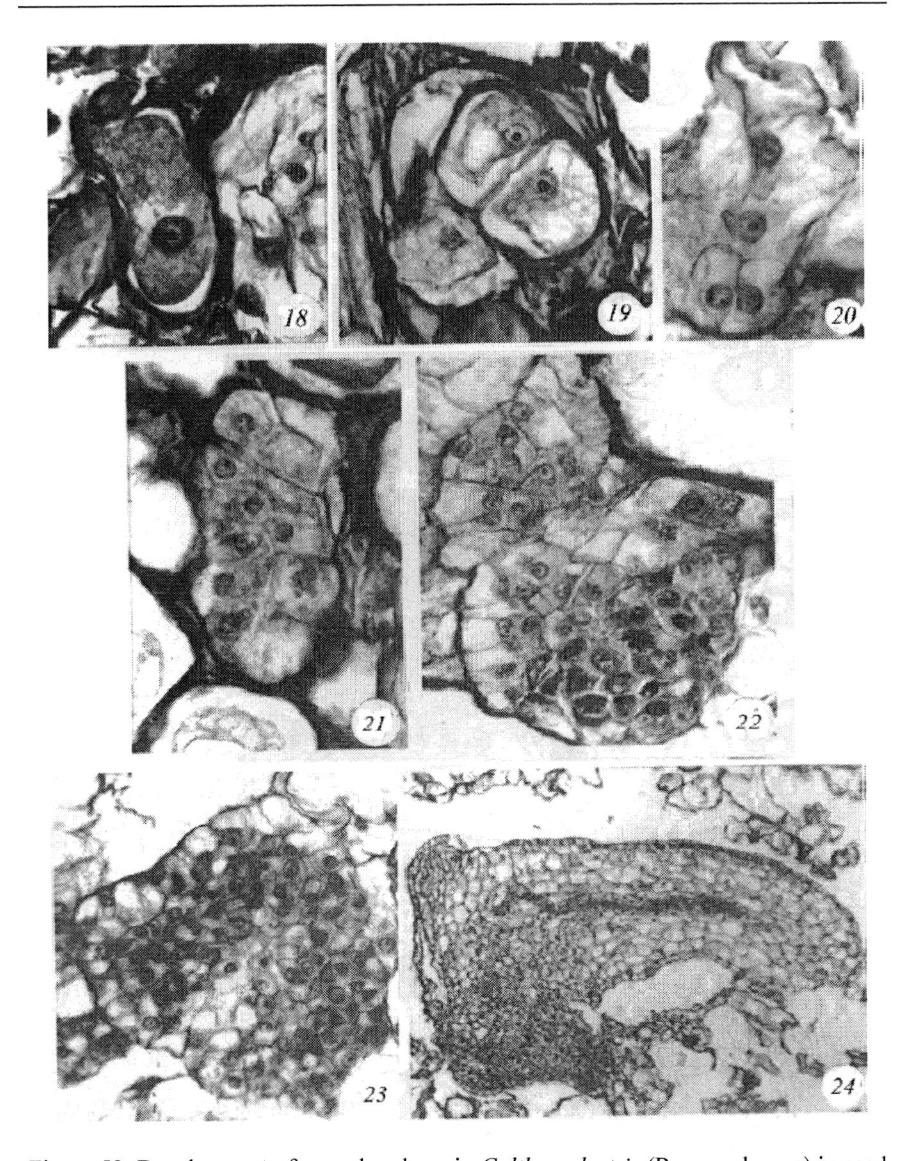

Figure 53. Development of sexual embryo in *Caltha palustris* (Ranunculaceae) in seed (*in vivo)* according to Onagrad-type (1–7); successive stages of development of embryoids *Aconitum heterophyllum* (Ranunculaceae) in culture *in vitro* according to Onagrad-type (8–15, 18–24), abnormalities can be found: early differentiation of xylem in the area of central cylinder (16, 17, 23) (Ahuja, Titova, Batygina, orig. data).

In the development of embryoid in *Aconitum heterophyllum* in callus culture *in vitro*, the clear resemblance with Onagrad-type of embryogenesis, typical for many others members of Ranunculaceae family, was discovered

(Figure 53). In the same time the different abnormalities, e.g., early differentiation of xylem elements in embryoids at heart-shaped stages could be observed. One of the causes of this is probably in the disturbance of differentiation at proembryo stage. The definite parallelism also was discovered in the formation of embryoid in callus and suspension cultures and sexual embryo *in vivo* in *Daucus carota.*

On the basis of information obtained, the following conclusion could be made. The resemblance in development of sexual embryos and embryoids manifests in the general regularities of morphogenesis: polarity (formation of a new axis), symmetry (radial and bilateral), cellular and histogenous differentiation, morphogenetic and morphophysiological correlations, and capacity for proliferation. There are critical stages in the genesis of both embryos: laying down of the first partitions, protoderm, differentiation of organs, autonomy, etc. In sexual and somatic embryo the polymorphism, the transitional forms between the embryo and the bud and some abnormalities are noted. Moreover, the development of sexual embryo is known to proceed according to the laws of cell division and embryogeny. However, *the variability of the first zygote division and the specifics of proceeding with further embryogenesis stages (8 types and more than 50 variations of embryogenesis)*, and the presence of transitional forms testify that *these rules and laws fail to be absolute*. A similar situation is observed also at the development of somatic embryos both in natural conditions and in culture *in vitro* (Batygina, 1996a, b).

Thus, the complex of morphogenetic features examined above testifies not only the parallelism in the development of sexual and somatic embryos but also the parallelism in the development of embryo and bud. It permits speaking about possible *evolutionary trends of transition from the bud to the "classical" sexual embryo (and/or back) through "midway forms": brood buds, embryoids, "non-classical" sexual embryos* (Figure 54). Perhaps, these structures *in the process of evolution have formed parallel and realized in any extent together with sexual embryo in different taxa.*

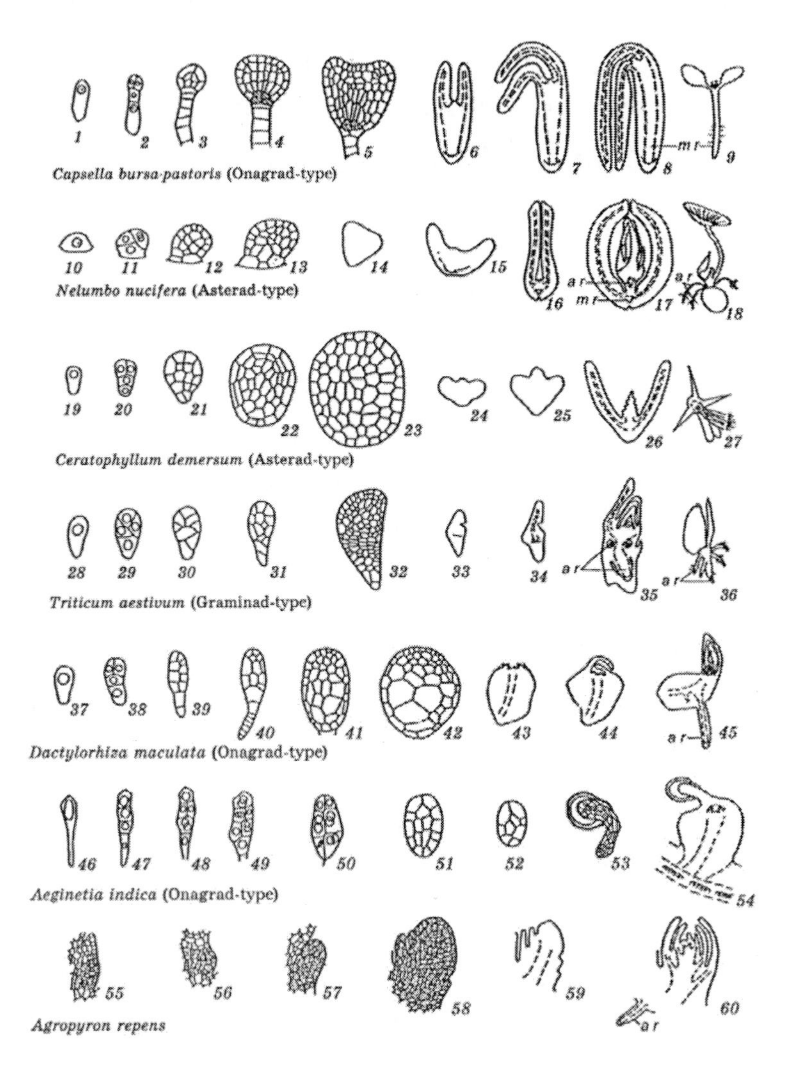

Figure 54. Parallelism of the first stages of morphogenesis of the sexual embryo and bud: (1–54) formation of the embryo and general view of the seedling:

(1–9) *Capsella bursa-pastoris* (mesophyte); (10–18) *Nelumbo nucifera* (hydrophyte), the main root is underdeveloped, adventive roots can be found; (19–27) *Ceratophyllum demersum* (hydatophyte), the main root is undeveloped; (28–36) *Triticum aestivum* (xerophyte), adventive roots can be found; (37–45) *Dactylorhiza maculata* (mycorhizic plant), there are no shoot and root apices, they appear only after germination; (46–54) *Aeginetia indica* (parasite plant), the embryo is reduced, the shoot apex appears only after germination, proper roots are not formed; (55–60) development of a lateral bud in *Agropyron repens*; shading shows hypophysial centre; *a r* – adventive root, *m r* – main root (from Batygina, Zakharova, 1997a).

Functioning of Stem Cells as a General Mechanism of Switching Over Developmental Programs in Ontogenesis

The reproduction systems of seed plants could be considered in relation to the character of genetic information transferring to posterity and modes of its realization. Under the influence of environmental factors not only the genetic information changes but also the modes of its transferring the mechanisms of which in many respects continue to be vague.

The study of the phenomenon "stem cells" is to be one of the prospective trends in the developmental biology aspect. The Department of Embryology and Reproductive Biology of the Komarov Botanical Institute of RAS has for many years been working out the general biological problem of the origin and role of stem cells in the ontogenesis of plant and animal organisms.

Preconditions of development of stem cells theory

The origins of the problem of stem cells (Maksimov, 1909) *in plants* are in the investigations of "*dormant meristem*" in a leaf (Sachs, 1887; Kerner von Marilaun, 1896, 1898; Naylor, 1932; Yarbrough, 1932; McVeigh, 1938; Batygina et al., 1995, 1996, 2006, 2010), ovule (Batygina, Freiberg, 1979; Batygina, 1991b; Batygina, Vinogradova, 2007), "*quiescent centre*" of root apex (Clowes, 1954), "*méristème d'attente*" in shoot apex (Buvat, 1955), *growing plants from somatic cells, haploidy, parasexual hybridization* (Vöchting, 1906; Haberlandt, 1902; Butenko, 1964; Reinert, 1963; see Plant tissue and cell culture, 1973; Haploids of higher plants *in vit*ro, 1986, etc.). For its solution the attracting of newest ideas and facts in the sphere of the study of animal and human stem cells is needed, discovered already in the beginning of XX century and obtaining a new resonance (Thomson, Marshal, 1998; Bosch, 2008; Lohmann, 2008) in relation with their great theoretical and practical significance.

The analysis of very first ontogenesis stages, namely, the specifics of its embryonic period in the members of numerous taxa of flowering plants *permitted arriving at the nontraditional conclusion (Batygina et al., 2004).*

Apical shoot and root meristems occur to be the product of functioning of zygote—the stem cell ("proximate stem cell" after Barlow, 1997). *The stem cells represent not a part of apical shoot and root meristems, as investigators traditionally considered* (Clark, 1997; Weigel, Jürgens, 2002; Groß-Hardt, Laux, 2003; Ivanov, 2003, 2007, etc.) *but its origin* (Figure 55) (Batygina et al., 2004; Batygina, Rudskiy, 2006; Batygina, Vinogradova, 2007; Batygina, 2007a-c, 2009, 2010, 2011)[2].

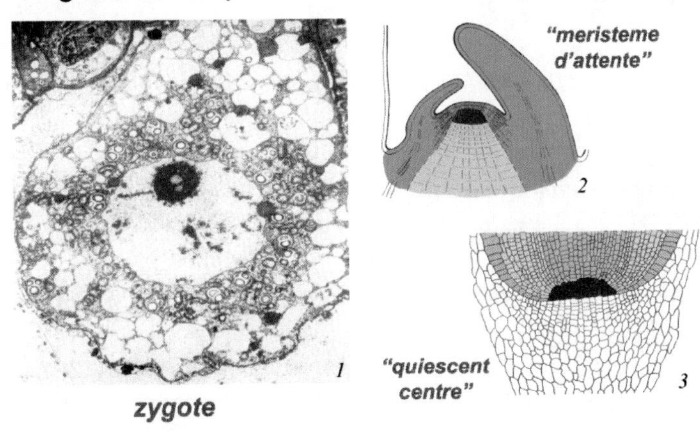

Figure 55. (1) zygote of *Hordeum vulgare* (from Norstog, 1972); (2) shoot apex of *Cheiranthus cheiri* (from Buvat, 1955); (3) root apex of *Zea mays* (from Clowes, 1954).

The study of reproductive biology, especially comparative embryology of members of flowering plants' different taxa allows considering that the stem cells form not only in shoot and root apices (méristème d'attente and quiescent centre). *The formation of stem cells derived from zygote is typical for all organs (flower, stem, leaf, root) and all stages of life cycle (sporophyte, gametophyte), moreover their functioning firstly depend on their localization and purpose.* It is the aspect of stem cells in which *initial cells of somatic embryos (embryoids) of different origin in natural conditions* and in culture *in vitro* should be examined. In relation with this the study of a new category of vegetative propagation—embryoidogeny—obtains a great significance (Batygina 1989a, b, 1993a, 2005a, b, 2009a, b, 2010, 2011; Batygina et al., 1979, 2004).

---

[2] I assume that egg cell and zygote in embryo sac, and the formation of sexual or parthenogenetic embryo, as well as nucellar, integumentary and cleavage embryoids are likely correctly to be considered as a result of cell lines producing from the first stages of ovule development. Owing to this, egg cell and zygote are believed to be the product of functioning of these cell lines.

Plasticity in development and reproduction of plants is firstly related with multisided activity of cells of plant body possessing the property of stem cells. According to the analysis of literature (Pottten, Loeffler, 1990; Clark, 1997; Barlow, 1997; Weigel, Jürgens, 2002; Ivanov, 2003, 2007; Byrne et al., 2003; Bosch, 2008; Lohmann, 2008) and original data the main *properties of plant stem cells* are revealed and postulated:

1. *toti- or pluripotency*, i.e., the capacity for formation of *not only different tissue and organ types but a new individual as well*, by various morphogenesis pathways (*embryo-, embryoido-, gemmorhizogenesis*);
2. *selfmaintenance, i.e., the creation of cell pool generally owing to symmetrical divisions and the system of intercellular interactions*;
3. *the capacity to proliferation and formation of cells-ancestors* of different tissue types ("niches")—owing to *asymmetrical divisions* under the definite signals;
4. *pulsatory and multistage character of the formation in tissue or organ and capacity to switch over the developmental program* that is provided by various molecular-genetic mechanisms (Batygina et al., 2004; Batygina, Rudskiy, 2006; Batygina, Vinogradova, 2007; Batygina, 2005a, b, 2007a-c, 2010, 2011).

According to the series of stem cell features formulated above, we manifested the unity of morphogenetic and reproductive processes at all stages of plant ontogenesis and life cycle. It confirms the necessity of the study from the position of organism integrity.

In spite of structure diversity in shoots and roots, all higher plants are characterized by the presence of quiescent centre or domain of undifferentiated cells in apical meristem that could consist even of a single cell as it does in sporiparous plants. *Dormant centres of shoot and root in some flowering plants appear in embryogenesis during epiphysis and hypophysis cell formation* (Figure 56A); *their differentiation usually occurs in fifth cell generation and is realized by means of asymmetrical divisions.* As a result of this, the united structural domain of undifferentiated cells, i.e. the initiation of all further organogenesis processes by definition could be created by cells of *independent lines.* The diversity of embryogenesis types known also demonstrates the independence of arising of shoot and root organs. For example, at Graminad-type of embryo development (Batygina, 1969) in cereals, both the root and shoot tissues could arise in very different cell lines

(Figure 57) (Barlow, 1997; Batygina, 2005c). However, the cells of the quiescent centre heterogeneous by origin are considered to be the stem ones in the traditional narrow sense of this term.

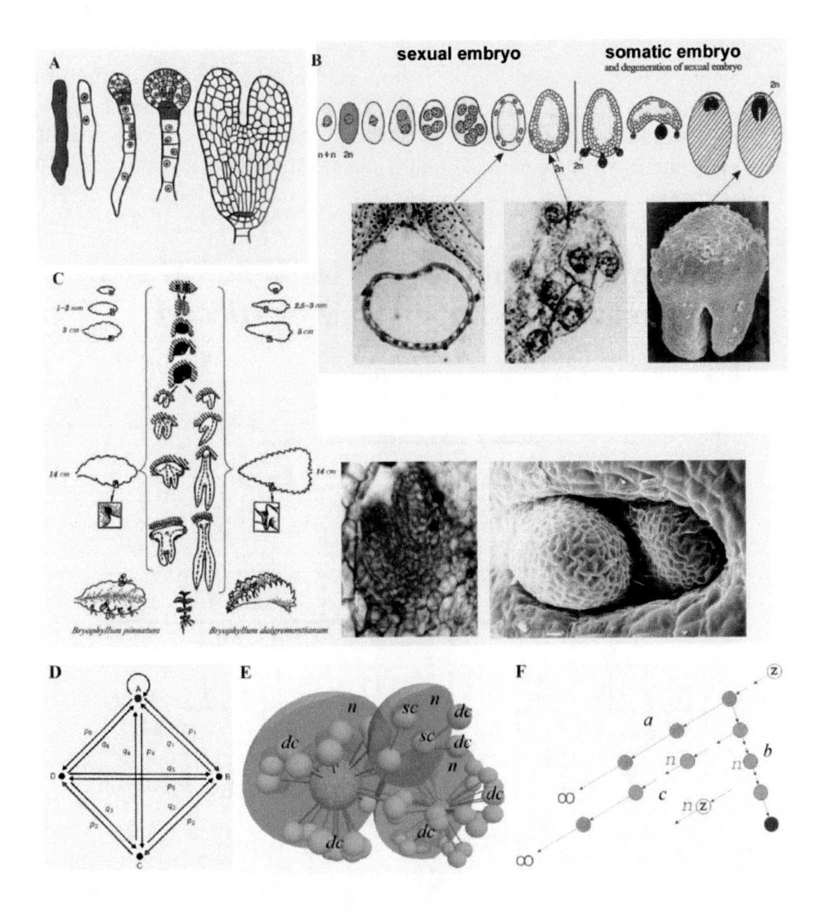

Figure 56. (A) embryogenesis *Arabidopsis thaliana.* There are stem cells, precursors of dormant centers of shoot and root apices.

(B) embryogenesis *Paeonia anomala.* Switching of development programs from heterophase to homophase one and formation of somatic embryos out of epidermis of the sexual embryo.

(C) formation of somatic embryos out of cells of "dormant" meristem in species of the genus *Bryophyllum.*

(D) basic morpho-phisiologycal states of a plant cell (Barlow, 1997). (A) stem cell, (B) meristematic cell, (C) differentiating cell, (D) mature diferentiated cell; A→B→C→D – most probable transitions.

Figure 56. (Continued).

(E) domains of stem cells and their niches (n – semi-transparent semi-spheres), which consist only of daughter cells (dc) cells, which differentiate in the initial of cell lines. In the surrounding (niche) of these cells sister cells (sc) can be found; these cells are at the same stage of differentiation, which principally distinguish them from the niche of the stem cells.

(F) three components of morphogenetic processes in plants: (a) potentially infinite selfmaintenace of stem cells; (b) differentiation and loss of stemness up to the death of a cell; (c) restoration of stemness and transdifferentiation in viable cells with undamaged genome, with zygote (Z) being one of them. (from Batygina, Rudskiy, 2006).

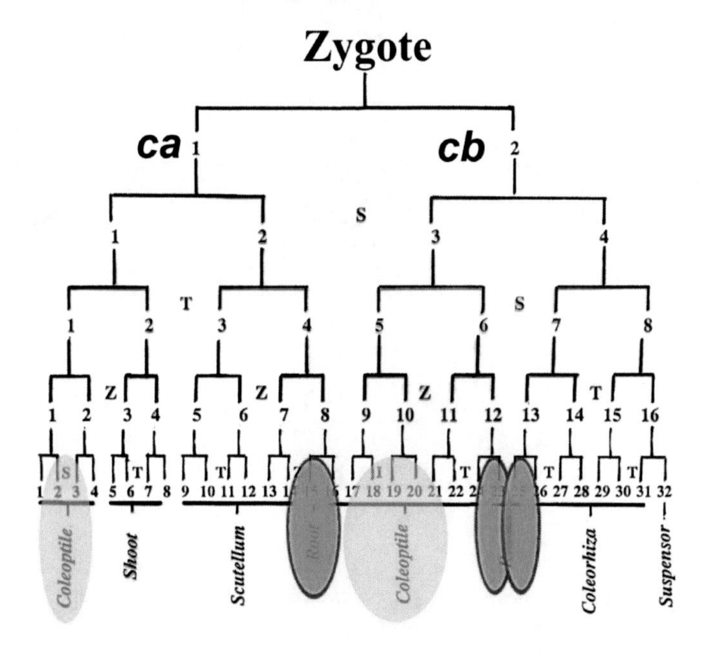

Figure 57. Cells of different lines can participate in creating of one embryo's organ. (from Barlow, Batygina, 2005).

The notion of stem cell, both in botanical literature and in zoological and medical one, we propose to be the ***morpho***-functional notion, because no universal genetic or epigenetic markers are known for identification of stem cells (Blau et al., 2001; Theise, Krause, 2002; Batygina, Rudskiy, 2006).

The complex WUS/CLV could be used as a marker for the shoot stem cells in the example of *Arabidopsis* genus (Veit, 2004), but it is not believed to be universal for the stem cells of a whole plant, in particular for the root where the products of the other genes fulfill the analogous function (Nakajima, Benfey, 2002). In one of the latest works on the model investigation of the role of CLV and WUS family proteins in regulation of shoot apical meristem structure the mechanism of maintenance of compartmental structure (space arrangement of zones of enzyme-marker synthesis) of shoot apical meristem was examined and the model of such mechanism was constructed (Figure 58) (Nikolaev et al., 2007).

Every plant cell is able to be in different morphophysiological states, the characteristics of one of them mostly correspond to the properties of stem cell, the others correspond to actively dividing meristematic cell. The transitions between these states have the stochastic character and in most cases histogenesis and organogenesis are connected with the loss of so called stemness but any other transitions are also probable (Barlow, 1997) (Figure 56D).

Besides the general space arrangement the stem cells are united by the processes proceeding *in domain.*

As we already stated above, relatively *symmetrical divisions are related to self-maintenance of the stem cell pool, the asymmetrical ones with transition of the daughter cell* to the definite pathway of differentiation. *The derivatives of stem cells, the initials of cell lines of tissues, create the niche* (Figure 56E). The surrounding of differentiated cell domains principally differs by the presence of sister cells (Figure 56E).

The zygote is not alone in its ability to be the initial cell of a plant. The diversity of morphogenetic processes comprehends *two components: potentially infinite meristematic activity* that is conditioned by the functioning of dormant centre cells, being in differentiated state (Figure 56D, A) and *the loss of stemness*— senescence and apoptosis (Figure 56D, B).

Such processes can take place at different ontogenesis stages and in all structures. For example, in *Paeonia embryoids* are formed from epidermal cells of sexual embryo (new sporophyte) (Batygina et al., 2004) (Figure 56B). The plasticity of leaf tissues (*Bryophyllum*, *Gnetum*) provides the possibility of the formation of *somatic embryos* from a single or group of cells (*dormant meristem=stem cells*), in other words, the repeating of early ontogenesis stages in the tissues of adult plant is observed (Batygina et al., 1996; Bragina et al., 2005) (Figure 56C). Programmed destruction of definite part of leaf primordium cells (*Anubias*) leads to the formation of epidermis owing to the transdifferentiation of interior tissues (Rudskiy et al., 2005, 2011).

The plasticity of generative structure cells of *maternal plant* (*Poa*) permits the formation of somatic embryos from ovule covers (nucellus or integument) (Batygina, 1991b). Thereby the cells of somatic embryos during the development use the niche of the sexual one. The phenomenon of gametophytic apomixis, consisting in the initiation of somatic embryo formation from egg cell or synergid—the cells of female gametophyte without fertilization, is widely spread among angiosperms (Czapik, 1997; Batygina, Vinogradova, 2007).

In anthers of cereals at definite developmental conditions, *the morphologically competent (stem) cells* of microspore transfer from usual *gametophytic* pathway (formation of pollen grains with sperms) to *sporophytic* one — formation of haploid regenerants.

At that, the cells of microsporangium walls are likely to be considered as a niche. In this case one of the most important properties of stem cells manifests itself, the possibility of repeated changing of developmental program.

In *in vitro* culture, the anther cells realize their potential by different morphogenesis pathways: *primary embryoidogenesis* (*secondary embryoidogenesis* is also possible), *gemmorhizogenesis,* gemmogenesis, rhizogenesis, histogenesis (Figures 33, 47) (Batygina, 1987a, 1989a, 2005b; Batygina et al., 1979).

The plasticity and polyvariety of these processes are conditioned by the diversity of the structure and origin of vegetative and generative structures in which the peculiar reserve of cells, possessing the property of stemness, is situated.

Hence, *in plant morphogenetic processes, even the third component is present—the restoring of stemness* in many respects depending on *stemness degree* (Figure 56F, c).

Thus, the plasticity and tolerance in plant development and reproduction is related firstly with multisided activity of stem cells at all stages of ontogenesis.

Special attention should be paid to *the unique property of plant cell—totipotency,* characterizing all morphogenetic possibilities (i.e., all the potential), that is habitual to individual and realized by different morphogenesis pathways. The final result of this property manifesting, its modes and forms of realization can be different and are conditioned by *the totipotency degree of the cell* (cells) (Batygina, 1987a).

SCHEME OF INTERACTION OF PRODUCTS
OF SHOOT APICAL MERISTEM CLV AND WUS GENES

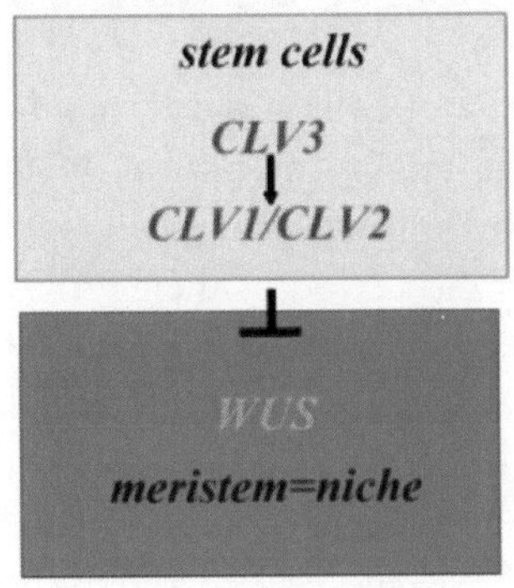

Figure 58. A certain balance between proliferation of stem cells and meristem cells in shoot apex of *Arabidopsis thaliana* is controlled by genes *CLV*1, *CLV*2, *CLV*3, etc. in stem cells, and *WUS* in surrounding niche cells – according to the principle of feedback. Imbalance leads to abnormalities in apex development (from Brand et al., 2000; Groß-Hardt, Laux, 2003; Weigel, Jürgens, 2002-2006).

Genetical control of meristem activity

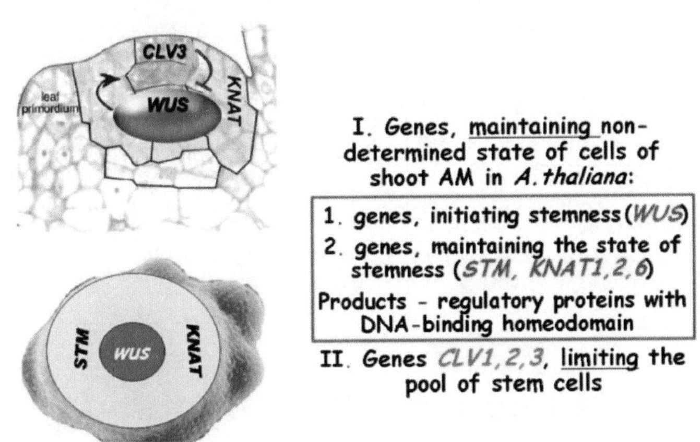

Figure 59. Time of meristems' life depends on work of regulatory genes, which suppress or activate expression of genes supporting non-determinated state of cells. It is possible to reveal such genes in model objects by study the mutants with alternated activity of meristems (from Ezhova et al., orig. data).

Not Every Cell Possesses Totipotency and Not Every Totipotent Cell Is Able to Become a Stem Cell

The degree of potency of cells – pluri-, omni- and toti-, is taxon-specific and is determined by the totality of factors, in the first turn by the system (tissue, organ, organism) from which the cell was obtained. The sexual cells (sperms and egg cell) in the period of their differentiation and specialization and more mature cells as well as the zygote in the period of its establishment and development are not belonged to totipotent ones. The property of totipotency recreates every time from the moment of zygote maturation and proliferation and losses by most embryo cells in the course of their specialization. However in some cases the definite cells of embryo, seedling and plant at different ontogenesis stages stay the stem ones – so called "méristème d'attente" or "dormant meristem", *that allows fulfilling the transition from one reproduction mode (heterophasic) into another (homophasic) and vice versa* (Figure 59).

In our opinion, *the phenomenon of multistage change of developmental program and creation of different subprograms in life cycle (probably repeated)* of most flowering plants is *universal*, and the *unique* property of plant cell — totipotency — lies in its base.

The direction of changes in developmental program, *the duration and the number of possible transitions are taxon specific* and in many respects *depend on external stress and internal* (hormonal, immunological, etc.) *factors*.

Thus, the changing in genetic program and polyvariety of transitions in plant *individual development* are *universal*. It is likely *conditioned by the alternation of of sexual and asexual modes of new individual formation in life cycle* that is conformed with G.P. Korotkova notions (1979) applying to animals.

The Phenomenon of Embryoidogeny and Its Role in Onto- and Phylogenesis

Comparative analysis of the results on the researches of the somatic embryo formation in natural conditions and in tissue culture permitted to formulate the notion of embryoidogeny phenomenon as a special type of vegetative propagation, and of embryoidogenesis as a process, in the base of which asexual mode of new sporophyte formation (i.e., without participation of gametes and fertilization) is laid. The propagation in this case is realized by either embryoids or miniature seedlings, developing on the plant from embryoids, or by the seeds, in which embryoids form (Batygina, 1987a, 2006a, 2009a etc.).

It is in conformity with the opinion of Winkler (1934), Battaglia (1963) and Grant (1981) that adventive embryony (formation of adventive embryos in seeds) and vegetative viviparity (on leaves, stems, etc.) (homophasic reproduction, T.B.) are the forms of vegetative propagation.

The phenomenon of embryoidogeny represented by different forms is of a greater adaptive significance for flowering plants than another type of vegetative propagation, gemmorhizogeny. During the evolution the origin of embryoids was likely conditioned by such their advantages as a shorter and consequently more power favourable pathway of the formation of new individuals' numerous initials with uniparental heredity (together with biparental embryos).

It is clear that the formation of different vegetative and generative diaspores, specially seeds containing the sporophyte initials of "new" daughter sexual generation (biparental heredity) and "old" maternal generation (uniparental heredity), i.e., the formation of agamosexual complexes obviously expands the possibilities of "struggle for existence", due to which these processes were picked up by natural selection and became the central item in preserving of biological diversity of plants.

One of the significant advantages of embryoidogenic propagation type is that the initial of a whole organism, forming in the seed or on the vegetative organs, occurs to be a unit of propagation. A full-value bipolar embryo, forming at embryoidogeny in some plant species is able not only to root immediately, but also disperses like seeds by wind or water over small and large distances, i.e. the area of dispersal increases, which is very important for preserving resources and biological diversity.

The peculiarity of such an original embryoidogenic type of vegetative reproduction (unlike the usual forms of vegetative propagation—particulation, sarmentation etc.) owes to the same features that are typical for propagation by seeds: the plurality of new individual initials and their capacity to close and far dispersion. Besides that, as it mentioned above, embryoidogeny extends essentially the diapason of seed heterogeneity. In consideration of polyembryony phenomenon and its different forms (monozygotic cleavage embryoidogeny, gametophytic apomixis, nucellar and integumentary embryoidogeny), sexual process, four categories of embryos and five types of genetic heterogeneity of seeds could be distinguished (Batygina, 1997, 1999a, b, 2000, 2006a, 2009a; Batygina, Vinogradova, 2007).

It should be mentioned once again that there are somatic embryos, forming in seeds together with sexual ones, that determine the type of heterogeneous seed. Apart from general seeds (only with the sexual embryo), there are seeds in which different modes of new individual formation—sexual and asexual—occur hand in hand.

It worth to be noted, that usually no one mode of reproduction is presented in pure form. For example, the sexual process precedes the arising of "monozygotic twins" (or triplets, etc.). In this case, one mode of new individual formation, sexual, is changed on another, asexual (change of morphogenetic developmental program).

Thus, in some cases, before "agamospermous seeds" in "clear state" were formed, i.e., containing only somatic embryos, the seeds in the process of their development were "agamosexual": they could contain nucellar, integumentary embryos and zygote, which subsequently could produce "monozygotic twins" by asexual mode—by the cloning (cleavage monozygotic embryoidogeny).

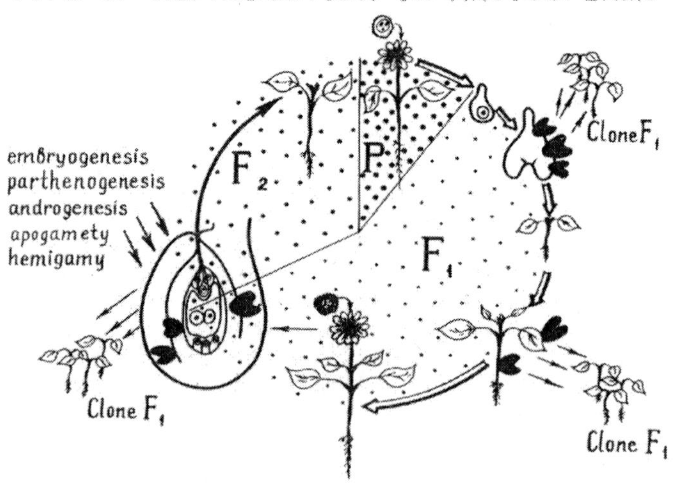

Figure 60. Embryoidogenic type of reproduction in angiosperm plants as a mechanism of appearing of the phenomenon of polyembryony and cloning of a plant organism at different stages of ontogenesis (from Batygina, 1993a).

Embryodogenic types of reproduction in angiosperms could be considered as the mechanism of the polyembryony phenomenon (*sensu lato*) arising and the cloning of plants at different ontogenesis stages (Figure 60).

All these complex *multistage processes* taking place in the seed ensure the capacity to *reproduce* and *propagate* new daughter individuals (or new generation) with different genotypes, either uniparental or biparental inheritance (in the case of sexual embryo formation).

Thus, the elaboration of a new conception of embryoidogeny phenomenon permitted the presentation *of a nontraditional view of the system of flowering plant* reproduction. It is in verification of the traditional and distinguishing of new notions connected with reproduction and propagation (see Chapter II).

There are *two modes of new individual formation — sexual and asexual; three pathways of morphogenesis in individual formation — embryogenesis, embryoidogenesis and gemmorhizogenesis; two forms of reproduction — heterophasic (embryogeny, gametophytic apomixis) and homophasic (reproductive and vegetative embryoidogeny, gemmorhizogeny — particulation, sarmentation and vegetative viviparity); two types of propagation — seed (gamospermy, agamospermy and gamo-agamospermy) and vegetative (which includes gemmorhizogenic and embryoidogenic modes of propagation).*

It worth to be noted especially, that the introduction of a new category of vegetative propagation—*embryoidogeny—allowed uniting such phenomena that earlier seemed to be isolated: monozygotic cleavage, nucellar and integumentary, gametophytic, foliar, cauligenous and rhizogenous embryoidogeny.* The question on the position of these phenomena in the reproduction system and their evolutionary role for many decades has been *debatable*.

According to above-mentioned, one is hard to agree with the statement of some authors, who attribute the special role to definite forms of embryoidogeny in the reproduction system of flowering plants. It is not quite correct, as for *determination of the role* of every embryoidogeny form *it must be considered from the comparative point of view* under interactions with the other forms and also with *embryogeny, gemmorhizogeny and gametophytic apomixis*.

In every plant cell all morphogenetic developmental information is laid down. During evolution the morphogenesis pathway and accordingly reproduction mode for every taxon have been fixed. Realization of the other morphogenetic potencies usually occurs in stress situation (hybridization, mutations, tissue culture, etc.) and is related with extraordinary properties of plant cell such as totipotency and stemness.

Analysis of a store of available data allowed to formulate the conclusion, that *namely the versatile activily of stem cells at all stages of plant organism ontogenesis causes the flexibility and tolerance of development of the organism and, correspondingly, of the plant reproduction systems.*

Chapter IV

Summary

- Stem cells in plants are formed in different organs (flower, stem, leaf, root) and at different stages of life cycle (sporophyte, gametophyte).
- The integrity of properties of stem cells—toti- and pluripotency, the ability for self-maintenance, pulsatory and multi-stage character of formation and especially the ability to switch over the morphogenetical developmental program—provides flexibility and tolerance of the reproduction system. It determines the functioning of a species-specific system of reliability (reserves and failures) of a plant at different stages of ontogenesis.
- The zygote—stem cell—is the "progenitor" of all stem cells of a plant.
- Stem cells are capable of forming a new individual, and various types of tissues and organs.
- Not every cell possesses the totipotency and not every totipotent cell is the stem cell, and not every stem cell is able to give new organism.
- The revealed role of stem cells gives new perspectives for

    - elaboration of reproduction theory and fundamental problems of developmental biology: the phenomenon of polyembryony, genetic heterogeneity of seeds, apomixis and others;
    - working out the problems of plant evolution: origination of flowering plants, gametophyte evolution, morphological nature of the ovule, embryo organs and others;
    - further development of innovations into biotechnology, selection and medicine.

Chapter V

Conclusion

Analysis of developmental biology of plants and animals allows to discuss some similarities in biology of their stem cells:

- structural–functional ensemble of integrated cell systems, including stem cells, provides a high degree of reliability of an individual:
  - the phenomenon of polyembryony
  - the phenomenon of viviparity
  - the phenomenon of parthenogenesis, androgenesis
  - the phenomenon of chimera formation
  - mitosis, meiosis, cell cycle, cytomixis, mutations, polytene chromosomes, fertilization, etc.

- activity of stem cells is realised through multivariability of:
  - modes of formation of an individual—sexual, asexual and parthenogenesis;
  - pathways of morphogenesis: sexual embryogenesis, somatic embryogenesis (embryoidogenesis[3]) and gemmogenesis (gemmorhizogenesis[3]);
  - morphological structural units of sexual and asexual reproduction.

- in the evolution of each taxon there has developed its own system of "hidden" reproductive reserves and failures – stem cells, working at different levels of hierarchy. It provides the flexibility and tolerance in

[3] - term used only for plants.

the development and reproduction of a plant at all stages of ontogenesis and population, on the whole

- the ability of stem cells of plants and animals to proliferate which determines their inducing, substituting, organizing and regulating role with close interaction with cells of the niche
- ability of the system of stem cells to switch the morphogenetic program in the life cycle of plants and animals is one of the mechanisms of transition from sexual to asexual reproduction and vice versa (for example, monozygotic twins)
- presence of multiple ontogeneses in the life cycle of plants and animals, which key mechanism is the stem cell
- the revealed main properties of stem cells of plants that are similar to those in animals

Not every cell possesses the totipotency and not every totipotent cell is able to become the stem cell, and not every stem cell is able to give new organism.

Chapter VI

Afterword

Perspective Approaches to the Problem of Stem Cells

The arsenal of the data received in the XX-XXI centuries on biology of development including stem cells biology has allowed me to propose and discuss the three main perspective trends in study of stem cells:

- *gametes – zygote – embryo – individual – population – coenosis;*
- *different types of somatic cells considering their localization and function in the organism; possibility of homologization of some somatic cells with the zygote;*
- *apical meristem of shoot apex and root apex in which organization somatic cells take part.*

I suggest a new priority trend in developmental biology: joint investigation and use of plant and animal stem cells for the purposes of humanity. The investigations of delicate points in the development and considering of all processes in dynamics is required in order to distinguish resemblance of plant and animal stem cells the presence of plasmalemma instead of hard cell wall (egg cell), the migration of cells and others.

– Use the system approach in the study of developmental biology and main principles of reproduction theory of plants and animals and, in particular, biology of their stem cells: architectonics, critical cell mass, rate of cell divisions, peculiarities of DNA and RNA synthesis considering the phase of cell cycle, etc.

- Extending the range of model objects, contrasting by systematic position, centers of origination, peculiarities of ecology of germination and ethology in population and coenosis, considering taxon-specific localization of the system of stem cells in the organism
- Comparison of patterns of morphogenesis and differentiation of reproductive and vegetative structures—in particular, detailed study of polymorphism and parallelism of development of sexual and somatic embryos (*in situ, in vivo,* and *in vitro*)
- Revealing of morphogenetic correlations and critical periods in morphogenesis
- Revealing of biological meaning of doubling structures (failures and reserves) in development and reproduction at different levels of hierarchy, e.g., systems of reliability
- It is necessary to discuss and create a glossary for researchers, involved in studying of various aspects of stem cells.
  - A special attention should be paid to the terms «totipotency», «pluripotency», «multipotency», «omnipotency», «unipotency» defining degree of cell's morphogenic potential.
  - In connection with various systems of stem cells and cell lines it is worth discussing the necessity of using the terms "degree of totipotency" and "degree of stemness" and "changes – loss and regain – of totipotency and stemness" in connection with the different character of development of cells, their differentiation, specialization and dedifferentiation (Batygina, 1978, 1987a-c, 2010, Batygina et al., 2004; Batygina, Rudskiy, 2006).

The perspectives mentioned above are the basis for the development of the theory of stem cell biology and will provide the opportunity to control separate stages of ontogenesis of plants and animals.

- Organizing the International Center of Reproduction and Stem Cells in Russia in the Komarov Botanical Institute of RAS in Saint Petersburg. The Research and Education Center for study of Embryology, Reproductive Biology, Reproductive Systems and Biotechnology has already been established on the basis of the Department of Embryology and Reproductive Biology, the leading

Russian school of development of reproduction theory, whose members are scientists from different countries around the world.

- Organizing the Council on Stem Cells and Reproduction with representatives of leading laboratories of Russia and different countries of the world.

Acknowledgments

The work was support by the Russian Academy of Sciences under the program "Biological resources of Russia: estimation of state and fundamental bases of monitoring" (2009-2011), the program of fundamental investigations of RAS "Biodiversity and dynamics of genofunds" (2006-2011); the program for support of leading scientific schools (grant no. NSh-7637.2010.4 "Elaboration of theory and revealing of reserves of plant reproduction. Morphogenetic programs and stem cells as a basis of stable development", 2010-2011); and Russian Foundation for Basic Researches (grant no. 11-04-01466 "Morphoprocesses at different modes of reproduction from positions of untraditional approaches and notions", 2011-2013).

I express the gratitude to the translator of the book I.V. Poutro.

A special appreciation to my colleague V.E. Vasilyeva for the help in preparation of manuscript. I also thankful to the colleagues who assisted me: O.G. Butuzova, E.A. Bragina and Ya.V. Osadchyi.

References

Afzelius K. Die Embryobildung bei *Nigritella nigra. Svensk. Bot. Tidskr.*, 1928, vol. 22, № 1-2, 82-91.

Barlow P.W. Stem cells and founder zones in plants, particularly their root. In: Poten C.S., editor. *Stem cells*. London; 1997; 29-57.

Barlow P.W. Embryos, stem cells and histogenetic compartments. Materials of III International School of Young Scientists «Embryology, Genetics and Biotechnology», Saratov, June 29 – July, 2009. Saratov: Publishing house of Saratov university, 2009; 16-17.

Barykina R.P. Sarmentation. In: Batygina T.B., editor *Embryology of Flowering Plants. Terminology and Concepts. Vol. 3. Reproductive systems.* St-Petersburg: Mir i Semya; 2000a; 302-305.

Barykina R.P. Particulation. In: Batygina T.B., editor. *Embryology of Flowering Plants. Terminology and Concepts. Vol. 3. Reproductive systems.* St-Petersburg: Mir i Semya; 2000b; 306-310.

Battaglia E. Apomixis. In: Maheshwari P., editor. *Recent Advances in the Embryology of Angiosperms*. Delhi: University of Delhi; 1963; 221-264.

Batygina T.B. Change of male gametes during fertilization in wheat. RAS SSSR, 1961a, Vol. 137, № 1, 220-223.

Batygina T.B. Some data on cytology of fertilization process in wheat. In: Plant Morphogenesis. Vol. 2. M., 1961б; 334-340.

Batygina T.B. On the possibility of separation of a new type of embryogenesis in Angiospermae, *Revue cytologie et biologie vegetale*. 1969, vol. 32, №2, 335-341.

Batygina T.B. Embryology of wheat. L: Kolos. 1974, 206 p.

Batygina T.B. Problems of morphogenesis in vivo and in vitro. Abstracts of Indian-Soviet Symposium "Embryology of Crop Plants", August 23-26, 1977. Leningrad, 1977, 41-42.

Batygina T.B. On certain regularities of morphogenesis at plant regeneration. Abstracts of region scientific conference "Theoretical questions of plant regeneration", October, 1978, Makhachkala. 1978. 13-14.

Batygina T.B. System approach to the problem of differentiation of angiosperms embryo // Materials of VIII All-Soviet Union Conference on Plant Embryology «Problems of Gametogenesis, Fertilization and Embryogenesis», October 12-13, 1982, Tashkent. Tashkent: Fan, 1983, 25-26.

Batygina T.B. Problems of morphogenesis *in situ*, *in vivo* and *in vitro*. Proceedings of the International Symposium "Plant tissue and cell culture application to crop improvement". 24-29 Sept. 1984. Olomouc, Chechoslovakia, Prague; 1984; 43-56.

Batygina T.B. The grain of cereals. Atlas. Leningrad: Nauka; 1987a.

Batygina T.B. New interpretation of asexual propagation and its classification in flower plants (in light of experimental data). Scientific reports of V All-Union School on theoretical plant morphology, Lvov; 1987b; 92-97.

Batygina T. B. New concept of asexual reproduction in flowering plants. Abstracts of XIV Int. Bot. Congress. Berlin; 1987c; p. 5.

Batygina T.B. A new approach to the system of reproduction in flowering plants. *Phytomorphology*, 1989a, vol. 39, № 1, 311-325.

Batygina T.B. New approach to the system of reproduction in flowering plants. *Apomixis Newslet*, 1989b, № 1, 52-55.

Batygina T.B. Embryoidogenic type – a new category of asexual reproduction in flowering plants. Proceedings of VII Intern. Congr. «Plant Tissue and Cell Culture», Amsterdam, 1990a, 244.

Batygina T.B. Embryoidogenic type of reproduction in flowering plants. *Apomixis Newsletter*, 1990b, № 2, 58-66.

Batygina T.B. New hypothesis about the initials and genesis of embryoids (somatic embryos) and a position of embryoidogeny in the reproduction system. *Apomixis Newsletter*, 1991a, № 3, 19-24.

Batygina T. B. Nucellar embryoidogeny in *Poa pratensis* (Poaceae). *Pol. Bot. Stud.*, 1991b, vol. 2, 121-125.

Batygina T. B. Position of the phenomenon of embryoidogeny in the system of flowering plants reproduction. Proceedings of XI Int. Symp. «Embryology and seed reproduction» Leningrad, 1990; St-Petersburg; 1992a; 6-10.

Batygina T.B. Embryological bases of plasticity and adaptive possibility of reproductive systems in flowering plants/ In: Yurtzsev B.A., editor. *Biological diversity: approaches to research and preservation.* St-Petersburg, Russia: Publishing House of Zoological Institute RAS; 1992b; 201-212.

Batygina T.B. Embryoidogeny – new category of flowering plant propagation mode. In: Teryokhin E.S., editor. Problems of reproductive biology of seed plants. *Transactions of Komarov Botanical Institute RAS.* Vol.8; 1993a; 15-25.

Batygina T.B. Certain aspects of morphogenetic polarity in plant ontogenesis. Abstracts of III congress of Russian Society of Plant Physiologists. St-Petersburg, 1993b. P. 258.

Batygina T.B. Anther as a model of research of morphogenetic potencies and morphogenesis pathways. In: Batygina T.B., editor. *Embryology of Flowering Plants. Terminology and Concepts. Vol. 1. Generative organs of flower.* St-Petersburg: Mir i Semya; 1994a; 120-121.

Batygina T.B. Apomixis, agamospermy and vivipary and their role in reproductive system of flowering plants. *Transactions of International symposium "Apomixis in plants: the state of problem and perspectives of studies"*; Saratov: Publishing House of Saratov State University; 1994b; 16-18.

Batygina T.B. Ovule and seed from the positions of biological system reliability. In: Batygina T.B., editor. *Embryology of Flowering Plants. Terminology and Concepts. Vol. 1. Generative organs of flower.* St-Petersburg: Mir i Semya; 1994c; 263-266.

Batygina T. B. Adaptive abilities of reproductive systems of flowering plants. Abstracts of Int. Conf. on Plants & Environmental Pollution, Lucknow, India ; 1996a ; p. 68.

Batygina T.B. Parallel development of somatic and sexual embryos. Abstracts of the 14 International Congress on Sexual Plant Reproduction, 18-23 February, 1996. Lorne, Australia; 1996b; p. 4.

Batygina T.B. Embryoidogeny. In: Batygina T.B., editor. *Embyology of Flowering Plants. Terminology and Concepts. Vol. 2. Seed.* St-Petersburg: Mir i Semya; 1997; 624-648.

Batygina T.B. Morphogenesis of somatic embryos developing in natural conditions. *Biologija*, 1998, № 3, 61-64.

Batygina T.B. Embryoidogenesis and morphogenesis of zygotic and somatic embryos. *Rus. J. Plant Physiol.*, 1999a, vol. 46, № 6, 884-898.

Batygina T.B. Genetic heterogeneity of seeds. *Acta Biol. Cracov. Ser. Bot.*, 1999b, vol. 41, № 1, 39-50.

Batygina T.B. Genetic heterogeneity of seeds. Plant physiology, 1999c, vol. 46, № 3, 438-453.

Batygina T.B. Embryoidogeny – new category of vegetative propagation. In: Batygina T.B., editor. *Embyology of Flowering Plants. Terminology and Concepts. Vol. 3. Reproductive systems.* St-Petersburg: Mir i Semya; 2000; 334-349.

Batygina T.B. Sexual and Asexual Processes in Reproductive Systems of Flowering Plants. *Acta Biol. Cracov. Ser. Bot*, 2005a, vol. 47, № 1, 51-60.

Batygina T.B. Theoretical bases of anther cultivation. In: Shamrov I.I., editor. *Embryological bases of wheat androcliny. Atlas.* Moscow: Nauka; 2005b; 9-23.

Batygina T.B. The theory of critical periods in flowering plants ontogenesis. Abstracts of the XVII International Botanical Congress. 17-23 July 2005, Austria, Vienna, 2005c. P. 312.

Batygina T.B. Embryoidogeny. In: Batygina T.B., editor *Embyology of Flowering Plants. Terminology and Concepts. Vol. 2. Seed.* Enfield (NH), Plymouth, UK: Science Publishers; 2006a; 403-409.

Batygina T.B. Polymorphism of sexual and somatic embryos as manifestation of their developmental parallelism under natural conditions and in tissue culture. In: Batygina T.B., editor *Embyology of Flowering Plants. Terminology and Concepts. Vol. 2. Seed.* Enfield (NH), Plymouth, UK: Science Publishers; 2006b; 409-419.

Batygina T.B. Structure of reproductive organs and reproduction of plants. In: Serebryakova T.I., Voronin N.S., Elenevskiy A.G., Batygina T.B., Shorina N.I., Savinych N.P. *Botany with basics of phytocenology: Anatomy and morphology of plants. Textbook for colleges.* Moscow: IKC Akademkniga, 2006c, 367-484.

Batygina T.B. Algorithms of morphogenesis. Polyembryony, gametophytic apomixis from position of stem cells. Abstracts of international symposium "Cellular, molecular and evolution aspects of morphogenesis. Moscow; 2007a. 15-18.

Batygina T.B. New algorithms of morphogenesis. VI congress of society of Russian plant physiologists. Materials of reports of international conference "Modern physiology of plants: from molecules to ecosystems", Syktyvkar, Komi. Part I. 2007b. 10-12.

Batygina T.B. Polyvariety of development of gametophyte and double structures as a base of polyembryony and genetic heterogeneity of seeds. Materials of II International School for young scientists "Embryology, genetics and biotechnology". Ufa: Publishing House of Bashkir state university; 2007c. 20-22.

Batygina T.B. Embryoidogeny is a new type of vegetative propagation. In: Batygina T.B., editor *Embryology of Flowering Plants. Terminology and Concepts. Vol. 3. Reproductive systems.* Enfield (NH), Plymouth, UK: Science Publishers; 2009a.

Batygina T.B. Phenomenon of embryoidogeny from the position of stem cells and its role in evolution. Materials of III International School of Young Scientists «Embryology, Genetics and Biotechnology», Saratov, June 29 – July, 2009. Saratov: Publishing house of Saratov university, 2009b. 17-21.

Batygina T.B. Theoretical bases of plant reproduction. In: Batygina T.B.; Kruglova N.N.; Gorbunova V.Yu.; Titova G.E.; Seldimirova O.A. From microspore to sort. Moscow: Nauka. 2010. 12-74.

Batygina T.B. Prerequisites of the conception of evolutionary developmental biology. Materials of the Conference "Morphogenesis in Individual and Historical Development", Moscow, March 16-18, 2011. Moscow: Publishing house of PIN, 2011. 11-13.

Batygina T.B.; Freiberg T.E. Polyembryony in *Poa pratensis* L. (Poaceae). *Botanical Journal,* 1979, vol. 64, №6, 793-804.

Batygina T.B.; Mametyeva T.B. On embryology of the genus *Poa* L. In: Yakovlev M.S., editor. Actual questions of angiosperm embryology. L: Nauka, 1979. 89-95.

Batygina T.B.; Rudskiy I.V. Role of stem cells in plant morphogenesis. *Reports of Russian Academy of Sciences,* 2006, vol. 410, № 5, 1-3.

Batygina T.B.; Vasilyeva V.E. Expediency of system approach to the problem of angiosperm embryo differentiation. *Ontogenesis*, 1983, vol. 14, № 3. 304-311.

Batygina T.B.; Valisyeva V.E. Plant reproduction. St-Petersburg: Publishing House of St-Petersburg State University; 2002, 230 p.

Batygina T.B.; Vasilyeva V.E. Periodization of development of reproductive structures. Critical periods. *Acta Biol. Cracov. Ser. Bot*, 2003, vol. 45, № 1, 27-36.

Batygina T.B., Vinogradova G.Yu. Polyembryony phenomenon. Genetic heterogeneity of seeds. *Ontogenesis*, 2007, vol. 38, № 3. 166-191.

Batygina T.B., Titova G.E., Rudskiy I.V., Bragina E.A. Embryo morphogenesis – one of plant ontogenesis model. Book of Abstracts of the

International Conference "Mathematical Models and Methods in Biology and Medicine", 29 May – 3 June, 2005. Bedlewo, Poland, 2005, p.17.

Batygina T.B.; Titova G.E.; Shamrov I.I., Bragina E.A., Vasilyeva V.E., Rudskiy I.V. Plant stem cell in term of embryology. *Acta Biol. Cracov. Ser. Bot*, 2005, Vol. 45, Supply 1, p. 51.

Batygina T.B.; Vinogradova G.Yu. Phenomenon of polyembryony. Genetic heterogeneity of seeds. *Ontogenesis*, 2007, Vol. 38, № 3, 166-191.

Batygina T.B.; Zakharova A.A. Polymorphism of sexual and somatic embryos as a evidence of their resemblance. *Bull. Pol. Acad. Sci. Biol. Sci*, 1997a, vol. 45, № 2-4, 235-255.

Batygina T.B.; Zakharova A.A. Parallels in the development of sexual and somatic embryos. In: Batygina T.B., editor. *Embryology of Flowering Plants. Terminology and Concepts. Vol. 2. Seed.* St-Petersburg: Mir i Semya; 1997b; 635-648.

Batygina T.B.; Bragina E.A.; Titova G.E. Morphogenesis of propagules in viviparous species *Bryophyllum daigremontianum* and *B. calycinum*. *Acta Soc. Bot. Polon.*, 1996, vol. 65, № 1-2, 127-133.

Batygina T.B.; Titova G.E.; Vasilyeva V.E. Plant reproduction: theoretical elaborations and innovation technologies. Innovations, 2007, №2 (100) 41-48.

Batygina T.B.; Vasilyeva V.E.; Mametyava T.B. Problems of morphogenesis *in vivo* and *in vitro* (embryoidogenesis in angiosperms). *Botanical Journal,* 1978, vol. 63, № 1, 87-111.

Batygina T.B.; Titova G.E.; Shamrov I.I.; Bragina E.A.; Vasilyeva V.E.; Rudskiy I.V. Problem of stem cells in plants (from position of embryology). Materials of X school on theoretical plant morphology "Constructional units in plant morphology". Kirov: Publishing House of Vyatka state university; 2004; 20-30.

Batygina T.B., Bragina E.A., Ereskovsky A.V., Ostrovsky A.N. Viviparity in plants and animals: invertebrates and lower chordates. PH of St.Petersburg university; 2006. P. 132.

Batygina T.B.; Kruglova N.N.; Gorbunova V.Yu.; Titova G.E.; Seldimirova O.A. From microspore to variety. Moscow: Nauka; 2010. 174 p.

Beklemishev V.N. On the general principles of life organization. *Bulletin of Moscow Society of Nature Experiments,* 1964, vol. 69, 22-38.

Beklemishev V.N. Methodology and systematics. Moscow: KMK Scientific press LTD; 1994.

Blau H.M.; Braselton T.R.; Weimann J.M. The evolving concept of a stem cell: entity or function? *Cell*, 2001, №105, 829-841.

Bosch T.G. Stem cells – from hydra to man.; Springer; 2008.

Bragina E.A.; Titova G.E.; Batygina T.B. The models of the formation of new sporophyte in genus *Bryophyllum* (Crassulaceae). Abstracts of VII Conf. of Plant Embryologists of Czech Republic, Slovakia and Poland. Lublin; 1995, p. 8.

Bragina E.A.; Arnautov N.N.; Batygina T.B. The phenomenon of vivipary in gymnosperms and angiosperms. Abstracts of the XVII International Botanical Congress. 17-23 July 2005, Austria, Vienna; 2005, p. 287.

Brandt U., Fletcher J.C., Hobe M., Meyerowitz E.M., Simon R. Dependence of stem cell fate in Arabidopsis on a feedback loop regulated by *CLV3* activity. *Science*, 2000, vol. 289, N 5479, 617-619.

Brien P. La croissance des spongillidae. Formation des choanocytes et des spicules. *Bull. Biol. France Belg.*, 1956, 110, 211-252.

Brukhin V.V.; Batygina T.B. Embryo culture and somatic embryogenesis in culture of *Paeonia anomala. Phytomorphology*, 1994, vol. 44, № (3-4), 151-157.

Butenko R.G. Culture of isolated tissues and physiology of plant morphogenesis. Moscow, Nauka, 1964. 272p.

Butenko R.G. Totipotency of plant cell and tissue culture. Transactions of the I All-Union conference "Culture of isolated plant organs, tissues and cells". Moscow, Nauka, 1970, 84-92.

Button J.; Kochba J.; Borman C.H. Fine structure of an embryoid development from embryogenic ovular callus of "Shamouti" orange (*Citrus sinensis* Osb.). *J. Exp. Bot.*, 1974, vol. 25, №85, 446-457.

Butuzova O.G., Pozdova L.M., Batygina T.B. The experimental data on the nature of seed dormance in various *Paeonia* species. Materials of III International School of Young Scientists «Embryology, Genetics and Biotechnology», Saratov, June 29 – July, 2009. Saratov: Publishing house of Saratov university, 2009; 26-28.

Buvat R. Le méristème apical de la tige. *Ann. Biol. Ser.*, 1955, vol. 31, 595-656.

Byrne M.E.; Kidner C.A.; Martienssen R.A. Plant stem cells: divergent pathways and common themes in shoots and roots. *Current Opinion in Genetics and Development*, 2003, vol. 13, Issue 5, № 10, 551-557.

Carmichael J.S., Friedman W.E. Double fertilization in *Gnetum gnemon*: The relationship between the cell cycle and sexual reproduction, *Plant cell*, 1995, 7, 1975-1988.

Clark S.E. Organ formation at the vegetative shoot meristem. *Plant Cell*, 1997, vol. 9, № 7, 1067-1076.

Clowes F.A.L. The promeristems and the minimal constructional center in grass root apices. *New Phytol.*, 1954, vol. 53, 108-116.

Cook R. A note on embryo rejuvenation. *Jour. Hered.*, 1938, 29, 419-422.

Crété P. La polyembryonie chez le *Lobelia syphiliyica* L. *Bull. Soc. Bot. France.*, 1938, vol. 85, № 7/8, 580-538.

Czapik R. Theoretical aspects of apogamety in angiosperms. *Bull. Pol. Acad. Sci., Biol. Sci.*, 1997, vol. 45, №2-4, 57-64.

Dogra P.D. The Embryology, Breeding Systems and Sterility in Cupressaceae – A Monograph; 1984; V. VI; 1-126.

Erdelská O. Polyembryony in maize – histological analysis. *Polish J. Bot.*, 1996, vol. 65, №1-2, 123-124.

Ernst A. Bastardierung als Ursache der Apogamie im Pflanzenreich. Eine Hypothese zur experimentellen Vererbungus-und Abstammungslehre. Jena: G. Fischer; 1918.

Ezhova T.A. Genetical control of flower morphogenesis. In: Developmental biolgy: morphogenesis of reproductive structures and role of somatic and stem cells in ontogenesis and evolution. Materials of international conference devoted to 50-year anniversary of the Laboratory of embryology and reproductive biology (Saint-Petersburg, December 13-16, 2010), Saint-Petersburg, 2010; 52-55.

Fagerlind F. Sporogenesis, Embryosackentwiclung und pseudogame Samenbildung bei *Rudbeckia laciniata*.*/ Acta Horti. Berg.*, 1946, 14(3), 39-90.

Fischer Ch.; Speth V.; Fleig-Eberenz S.; Neuhaus G. Induction of zygotic polyembryos in wheat: influence of auxin polar transport. *The Plant Cell*, 1997, vol. 9, 1767-1780.

Frost H.B. Nucellar embryony and juvenile characters in clonal varieties of *Citrus*. *Jour. Hered.*, 1938, vol. 29, 423-432.

Gamborg O.L.; Miller R.A.; Ojima K. Nutrient requirements of suspension cultures of soybean root cells. *Exp. Cell. Res.*, 1968, vol. 50, 151-158.

Garcês H.M.P.; Champagne C.E.M.; Townsley B.T.; Park S.; Malhó R.; Pedroso M.C.; Harada J.J.; Sinha N.R. Evolution of asexual reproduction in leaves of the genus *Kalanchoë*. *PNAS.*, 2007, vol. 104, №39, 15578-15583.

Gerassimova-Navashina E.N. The mitotic hypothesis of double fertilization, *Rept. Acad. Sci. USSR*, 1947, 57, 395-492.

Gerassimova-Navashina E.N. On the gametophyte and main features of development and functioning of reproducing elements in the angiosperms. *Problems of Botany*, 1958, 3, 125-167.

Gerassimova-Navashina H.N. A contribution to the cytology of fertilization in flowering plants, *Nucleus*, 1960, 3, 111-120.

Gerassimova-Navashina H. Some cytological aspects of double fertilization. *Rev. Cytol. Végétales*, 1969, 32(3-4), 301-308.

Gerassimova-Navashina H.N. On the periods of mitotic cycle of gametes in angiosperms, *Bull. Moscow State Univ.*, Ser. Biol., 1971a, 3, 93-94.

Gerassimova-Navashina E.N. Process of double fertilization in angiosperms and mitotic cycle of the cell, *Phytomorphology*, 1982, 32, 222-233.

Grant V. Population structure in relation to macroevolution. *Biol. Zentralblatt*, 1977, Bd. 96, 129-139.

Grant V. Plant Speciation. N.-Y.: Columbia Univ. Press; 1981.

Groβ-Hardt R.; Laux T. Stem cell regulation in the shoot meristem. *J. Cell Sci.*, 2003, vol. 116, № 9, 1659-1666.

Haberlandt G. Kulturversuche mit isolierten Pflanzenzellen. *Sitzungsber. Akad. Wiss. Wien, Math. nat.*, 1902, K 1, III, p. 69-92.

Haccius B. Weitere Untersuchungen über Somatogenese aus den Suspensorenzellen von *Eranthis hiemalis* Embryonen. *Planta*, 1965, vol. 64, № 3, 219-224.

Haccius B. Question of unicellular origin of non-zygotic embryos in callus cultures. *Phytomorphology*, 1978, vol. 28, 74-81.

Hall J.G. An embryological study of *Limnocharis emarginata*. *Bot. Gaz.*, 1902, № 33, 214-219.

Hanstein J. Die Entwicklung des Keimes der Monokotylen und Dikotylen // Bot. Abh. 1870. Bd 1. P. 1-122.

Higashiyama T., Kuroiwa H., Kawano S., Kuroiwa T. Kinetics of double fertilization in Torenia fournieri based on direct observations of the naked embryo sac. *Planta*, 1997, 203, 101-110.

Hodgson R.W.; Cameron S.H. Effects of reproduction by nucellar embryony on clonal characteristics in *Citrus*. *Jour. Hered.*, 1938, vol. 29, 417-419.

Hofmeister W. Vergleichende Untersuchungen der Keimung, Entfaltung und Fruchtbildung der höherer Kryptogamen und der Samenbildung der Coniferen. Leipzig: Verlag von Friedrich Hofmeister; 1851.

Howe M.D. A morphological study of the leaf notches of *Bryophyllum calycinum*. *Amer. J. Bot.*, 1931, vol. 18, № 5, 387-390.

Ilyina G.M. Comparative embryological study of poppies family (Papaveraceae Juss.) in connection with its position in angiosperm system. PHd theses. Moscow; 1968.

Ivanov V.B. Problem of stem cells in plants. *Ontogenesis,* 2003, vol. 34, № 4, 253-261.

Ivanov V.B. Stem cells in the root and the problem of stem cells in plants. *Ontogenesis,* 2007, vol. 38, № 6, 406-419.

Ivanova-Kazas O.M. Asexual propagation. Leningrad: Publishing House of Leningrad State University; 1977, 240p.

Jeffrey E.C. Polyembryony in *Erythronium americanum. Ann. Bot*., 1895, № 9, 537-541.

Johansen D.A. Plant embryology. Embryogeny of the Spermatophyta. Waltham;1950.

Johnson M. A. The origin of the pseudo-bulbils *Kalanchoë daigremontiana. Bull. Torrey Bot. Club*., 1934, vol. 61, 355-366.

Kamelina O.P. Possibility of Distinguishing a Tubifloral Type of Endosperm Development. In: Batygina T.B., editor. *Embryology of Flowering Plants. Terminology and Concepts. Vol. 2. Seed.* St-Petersburg: Mir i Semya; 1997; 281-283.

Kanaev I.I. Twins and genetics. Leningrad: Nauka; 1968.

Kerner von Marilaun A. Pflanzenleben, Vol.1, Verlag Bibliographische Instituts, Leipzig-Wien, 1896, 766p.

Kerner von Marilaun A. Pflanzenleben, Vol.2, Verlag Bibliographische Instituts, Leipzig-Wien, 1898, 778p.

Kerstetter R.A., Hake S. Shoot meristem formation in vegetative development. *Plant Cell*, 1997, vol. 9, 1001-1010.

Khan R. Contributions to the morphology of *Ephedra foliate* Bioss. II. Fertilization and Embryogeny // *Proc. Nat. Acad. Sci. Ind*., 1943, vol. 13, № 4, 357-375.

Khokhlov S.S. General issues of haploidy, Haploidy and breeding, Moscow, Nauka, 1976, 5-14.

Kolesova G.E.; Batygina T.B. Nelumbonaceae family. In: *Comparative anatomy of seeds.* Leningrad: Nauka; 1988; vol. 2; 157-162.

Koltsov N.K. Role of a gene in developmental. *Biological journal*, 1935, vol. 4(5), pp. 753-774.

Konar R.N.; Nataraja K. Experimental studies in *Ranunculus sceleratus* L. Development of embryos from the stem epidermis. *Phytomorphology*, 1965, vol. 15, №2, 132-137.

Konar R. N.; Thomas E.; Street H. E. Origin and structure of embryoids arising from epidermal cells of the stem of *Ranunculus sceleratus* L. *J. Cell Sci*., 1972, vol. 11, 77-93.

Korotkova G.P. Origin and evolution of ontogenesis. Leningrad: Publishing House of Leningrad State University; 1979.

Korotkova G.P., Tokin B.P. Embryology and genetics. Disputable questions. Leningrad: Publishing House of Leningrad State University; 1977, 63 p.

Kursanov L.I.; Krasheninnikov F.I.; Komarnitskiy N.A. and Kursanov A.L. Handbook on Botany for the Higher Pedagogical Educational Establishments and Universities, vol. 1. Gosudarstvennoe Uchebno-Pedagogicheskoe Izdatelstvo Narkomprosa RSFSR, Moskow, Russia, 1940;

Land W.J. Spermatogenesis and oogenesis in *Ephedra trifurcate // Bot. Gaz.* 1904. Vol. 38, № 1, P. 1-18.

Levina R.P. Reproductive biology of seed plants. Review of the problem. Moscow: Nauka; 1981.

Linnaeus C. Flora lapponica, Amstelaedami, 1737.

Lobashov M.E. Genetics. 2-nd edition. Leningrad: Publishing House of Leningrad State University, 1969, 752p.

Lohmann J.U. Plant Stem Cells: Divide et Impera. In Bosch T.G., editor. *Stem cells – from hydra to man.* Springer; 2008, pp. 1-16.

Maheshwari P. An Introduction to the Embryology of Angiosperms. N.-Y.: McGraw-Hill Book Company; 1950.

Maximov A. Der Lymphozyt als gemeinsame Stammzelle der verschiedenen Blutelemente in der embryonalen Entwicklung und im postfetalen Leben der Säugetiere. Folia Haematologica, Leipzig, 1909, 125-134.

McVeigh J. Regeneration in *Crassula multicava. Amer. J. Bot.*, 1938, vol. 25, 7-11.

Modilewksy Y.S. On embryos of *Allium odorum* L. *Reports of Kiev Botanical Garden*, 1931, №(12-13). 27-48.

Mogensen H.L. Double fertilization in barley and cytological explanation for haploid embryo formation, embryoless caryopses, and ovule abortion, *Carlsberg Res. Commun.*, 1982, 47, 313-354.

Morgan T.G. Structural basics of heredity. Moscow, Leningrad, 1924, 310 p.

Morgan T.G. Development and heredity Развитие. М.Л., Moscow, Leningrad, 1937, 242 p.

Nakajima, K., Benfey, P.N. Signaling in and out: control of cell division and proliferation in the shoot and root. *The Plant Cell*, 2002, vol. 14 (Suppl.), 265-276.

Naumova T.N. Apomixis in angiosperms. Nucellar and integumental embryony. Boca Raton: CRC Press Inc., 1993. 144 p.

Naylor E.E. The morphology of regeneration in *Bryophyllum calycinum. Amer. J. Bot.*, 1932, vol. 19, № 1, 32-40.

Nikolaev S.V.; Penenko A.V.; Lavrekha V.V.; Melsness E.D.; Kolchanov L.A. Model investigation of CLV1, CLV2, CLV3 and WUS proteins in regulation of shoot apical meristem structure. *Ontogenesis,* 2007, vol. 38, № 6, 457-462.

Norstog K. Early development of the barley embryo: Fine structure. *Am. J. Bot.*, 1972, 59, 123-132.

Pisyaukova V.V. Elements of morphological plant evolution. Textbook. Leningrad Publishing House of Leningrad state pedagogical institute; 1980.

Pijl L. van der. Princeples of dispersal in higher plants. Berlin etc.: Springer-Verlag; 1969.

Poddubnaya-Arnoldi V.A. Polyembryony in the Orchids. *Bull. Main Bot. Garden USSR Acad. Sci*, 1960, vol. 36, 56-61.

Poddubnaya-Arnoldi V.A. Cytoembryology of angiosperms. Moscow: Nauka; 1976.

Potten C.S.; Loeffler M. Stem cells: attributes, cycles, spirals, pitfalls and uncertainties. Lessons for and from the crypt. *Development*, 1990, vol. 110, 1001-1020.

Raghavan V. Origin of the quiescent center in the root of *Capsella bursa-pastoris* (L.) Medik. *Planta,* 1990, vol. 181, 62-70.

Reinert J. Experimental modification of organogenesis in plant tissue cultures. In: Plant tissue and organ culture. Sympos. Int. Soc. Plant Morphologist, Delhi, 1963, 168-177.

Reinert J. Morphogenese in Gewebe- und Zellkulturen. Naturwiss., 55, 4, 1968.

Reinert J.; Bajaj Y.P.S. Plant cell, tissue, and organ culture. New York: Springer-Verlag, 1977.

Resende F. Variabilidade fenoe genotipicamente determinada na fisiologia do desenvolvimento dos pseudo-bolbilhos (*Bryophyllum*). *Bol. Soc Port. Ci. Nat.*, 1954, vol. 20, №1, 78.

Rodin A. Reflections about an art. Memoirs of contemporaries. Moscow: Respublika, 2000, 367p.

Rudskiy I.V.; Batygina T.B.; Titova G.E. Determination mechanisms of mature plant construction in embryogenesis and germination in three species of Araceae Juss. family. Abstracts of the XVII International Botanical Congress, 17-23 July 2005, Austria, Vienna; 2005; p. 305.

Rudskiy I.V.; Titova G.E.; Batygina T.B. Analysis of Space-Temporal Symmetry in the Early Embryogenesis of *Calla palustris* L., Araceae. *Math. Model. Nat. Phenom.,* 2011, vol. 6, № 2, 82-106.

Sachs J. On the physiology of plants. 1887.

Savina G.I. Fertilization in *Cypripedium calceolus* L. *Botanical Journal,* 1964, vol. 49, №9; 1317-1322.

Saxton W.T. Contributions to the life-history of *Actinostrobus pyramidalis. // Ann. Bot.* 1913. Vol. 27, P. 321-345.

Seldimirova O.A.; Titova G.E. Peculiarities of secondary embryogenesis in the culture *in vitro* of spring soft wheat anthers. Materials of II International School for young scientists "Embryology, genetics and biotechnology". Ufa: Publishing House of Bashkir state university; 2007. 105-106.

Shamrov I.I., Dyachuk T.I., Batygina T.B., Dyachuk P.A. Embryoidogenous type of asexual propagation and classification of anomalies in culture of anther on the example of wheat. Materials of International Conference "Biology of Cultivated Cells and Biotechnology", Novosibirsk, 1988, 210-211.

Shmalgausen I.I. Factors of evolution. Theory of stabilizing selection. Moscow: Nauka; 1968, 451p.

Singh H., Oberoi J.P. A contribution to the life history of *Biota orientalis* Endl. *Phytomorphology*. 1962. V. 12, №4, 373-393.

Smirnova O.V. Peculiarities of vegetative propagation of herbs in oak forests in connection with questions of self-maintenance of population. In: Age content of flowering plant populations in relation to their ontogenesis. Moscow: Publishing House of Moscow state pedagogical institute; 1974; 168-195.

Snegirevskaya N.S. Materials towards morphology and systematics of genus *Nelumbo* Adans. *Transactions of Komarov Botanical Institute of USSR AS,* 1964, vol. 1, №13, 104-172.

Souèges R. L'hypophyse et l'epiphyse: les problems d'histogenese qui leur sont lies. *Bull. Soc. Bot. France,* 1934, vol. 81, 737-748, 769-778.

Souèges R. Exposes d'embryologie et de morphologie végétales. VII. Les lois du dèveloppement. *Act. Sci. Industr.*, 1937, vol. 521, 1-94.

Souèges R. Exposes d'embryologie et de morphologie végétales. X. Embryogénie et classification. Deuxieme fascicule: Essai d'un systeme embryogénique (Partie générale). *Act. Sci. Industr.*, 1939, p. 85.

Stow I. Experiental studies on the formation of embryo sac-like giant pollen grain in the anther of *Hyacinthus orientalis*. *Cytologia*, 1930, 1, 417-439.

Stow I. On the female tendencies of the embryo sac-like giant pollen grain of *Hiacinthus orientalis*. *Cytologia*, 1934, 5, 88-108.

Swamy B.G.L. Gametogenesis and embryogeny of *Eulophea epidendrae* Fischer. *Proc. Natl. Inst. Sci., India.*, 1943, № 9, 59-65.

Swamy B.G.L. Agamospermy in *Spiranthes cernua. Lloydia*, 1948, vol.11, №3, 149-162.

Takhtajan A.L. Bases of evolutionary morphology of angiosperms. Moscow-Leningrad: Nauka; 1964.

Takhtajan A.L. Principles of organization and transformation of complex systems: system approach. St-Petersburg; 1998.

Teryokhin E.S. The metamorphosis of ontogeny of flowering parasitic plants. (In Russian). Reports of Acad. Sci. of USSR. 1968, 178. P. 857-859.

Teryokhin E.S. About possibility of using of notion "behavior" in plant evolution study. Bot. Jurn. 1972, Vol. 57, № 1, 75-89.

Teryokhin E.S. Parasitic flowering plants. Evolution of ontogenesis and mode of life. Leningrad: Nauka; 1977.

Teryokhin E.S. Problems of evolution seed plant ontogenesis. In: Petrov Yu.E., editor. *Transactions of Komarov Botanical Institute of USSR AS.* vol.2. 1991.

Teryokhin E.S. Seed and seed propagation. St-Petersburg: Mir i Semya; 1996, 376p.

Teryokhin E.S., Schuchardt B., Wegmann K. Types and forms of vegetative propagation in the Orobanchaceae as a result of different adaptive strategies. In: Moreno M.T. et al. (eds.) Advanced in parasitic plant research. Proceed. 6th Internat. Symposium. Cadoba. Spain. P. 243-248.

Theise N.D.; Krause D.S. Toward a new paradigm of cell plasticity. *Leukemia*, 2002, № 16, 542-548.

Thomson J.A., Marshal V.S. Primate embryonic stem cells. *Cur. Top Develop. Boil.*, 1998, vol. 38, 133-165.

Timofeev-Resovsky N.V. From the history of problem of interrelation of micro- and macroevolution. Materials of symposium "Micro- and macroevolution", Tartu, September 2-5, 1980. Tartu, 1980, 7-12.

Tyrnov V.S. Haploidy in plants: scientific and applied significance. Moscow: Nauka, 1998, 53p.

Upadhyay N., Makoveychuk A. Yu., Nikolaeva L.A.; Batygina T.B. Organogenesis and somatic embryogenesis in leaf callus culture of *Rauwolfia caffra* Sond. *Journal of Plant Physiology*, 1992, vol. 140, № 2. 218-222.

Vasilyeva V.E., Batygina T.B. Autonomy of Embryo In: Batygina T.B., editor *Embyology of Flowering Plants. Terminology and Concepts. Vol. 2. Seed.* Enfield (NH), Plymouth, UK: Science Publishers; 2006, 375-382.

Veit B. Determination of cell fate in apical meristems. *Curr. Opin. Plant Biol.*, 2004, vol. 7, 57-64.

Vöchting H. Über Transplantnation am Pflanzenkörper. Tübingen, 1892.

Vöchting H. Über Regeneration und Polrität bei höhern Pflanzen. Botanisches Zeitung, 1906, vol. 64, 101–148, plates V–VII.

Warden J. Cytological observation of first phases of induced leaf-plantlet development in *Bryophyllum crenatum*. Their relation to gene activation. *Rev. Biol.*, 1968, vol. 6, p 357.

Weigel D.; Jürgens G. Stem cells that make stems. *Nature*, 2002, vol. 415, 751-754.

White

Winkler H. Fortpflanzung der Gewächse. VII. Apomixis. Handwörterbuch der Naturwissenschaft. Jena, 1934, vol. 4, 451-461.

Yakovlev M.S. Paeoniaceae family. In: Yakovlev M.S., editor. *Comparative embryology of flowering plants. Phytolaccaceae-Thymelaeaceae*. Leningrad: Nauka; 1983. 70-77.

Yakovlev M.S.; Yoffe M.D. On some peculiar features in the embryogeny of *Paeonia* L. *Phytomorphology*, 1957, vol. 7, № 1, 74-82.

Yakovlev M.S.; Yoffe M.D. Further research of new type of angiosperm embryogenesis. *Botanical Journal,* 1961, vol. 46, №10, 1402-1421.

Yakovlev M.S., Zhukova G.Y. Angiosperms with green and colorless embryo (chloro- and leucoembryophytes). Nauka Press, Leningrad, 1973, 1-116.

Yarbrough J.A. Anatomical and developmental studies of the foliar embryos of *Bryophyllum calycinum. Amer. J. Bot.*, 1932, vol. 19, № 6, 443-453.

Yarbrough J.A. History of leaf development in *Bryophyllum calycinum. Amer. J. Bot.,* 1934, vol. 21, №8, 467-482.

Additional References

Andronova E.V. Epiphysis. In: Batygina T.B., editor. *Embyology of Flowering Plants. Terminology and Concepts. Vol. 2. Seed.* St-Petersburg: Mir i Semya; 1997; 356-358.

Astaurov B.L. Genetics and the problems of individual development. *Ontogenesis,* 1972, vol. 3. №6, 547-565

Batygin N.F. Ontogenesis of higher plants. In: *Nikolay Fyodorovitch Batygin. Life and work.* St-Petersburg: NPO Professional, 2007, 23-140.

Batygin N.F. System approach in biology and agronomy. In: *Nikolay Fyodorovitch Batygin. Life and work.* St-Petersburg: NPO Professional, 2007, 151-196.

Batygina T.B. Ovule and seed viewed from reliability of biological systems // Embryology of flowering plants. Terminology and concepts / Ed. Batygina T.B. Enfield (NH), USA-Plymouth, UK, 2002a. Vol. 1; 214-217.

Batygina T.B., Shamrov I.I. Ovule and seed structure in Poa, Paeonia and orchids from the point of reliability. Abstr. XV Intern. Bot. Congr. Yokohama, Japan, 1993, p. 96.

Batygina T.B.; Vasilyeva V.E. Sexual reproduction in flowering plants: periodization of egg cell and zygote development and possible types of caryogamy. In: B.Bhatia et al. (eds.), *Plant Form and Function*, Angkor Publishers, New Delhi, 1997, 170-198.

Batygina T.B.; Vasilyeva V.E. *In vivo* fertilization. In: Bhojwani S.S., Soh W.Y. (eds.), *Current Trends in the Embryology of Angiosperms.* Dordrecht, Boston, London: Kluwer Academic Publishers, 2001, 101-142.

Bhandari N.N.; Soman P.; Bhargava M. Histochemical studies on the female gametophyte of *Argemone mexicana* L. *Cytologia*, 1980, vol. 45, 1-2, 281-291.

Boris Lvovitch Astaurov: essays, memoirs, letters, materials. Moscow: Nauka, 2004, 427 p.

Bouman F. The ovule. In Johri B.M., editor. *Embryology of angiosperms*. Berlin ets.; 1984; 123-157.

Brossard D. Le bourgennement èpiphylle chez le *Bryophylllum daigremontianum* Berger (Crassulaceae). Étude cytochimique, cytophotometrique et ultrastrucurale. *Ann. Soc. Nat. Bot.*, 1973, 12 series, vol. 14, 93-214.

Bugara A.M. Cytochemistry of meiosis. In: Batygina T.B., editor. *Embryology of Flowering Plants. Terminology and Concepts. Vol. 1. Generative organs of flower*. St-Petersburg: Mir i Semya; 1994; 76-81.

Carmichael J.S., Friedman W.E. Double fertilization in *Gnetum gnemon* (*Gnetaceae*): Its bearing on the evolution of sexual reproduction within the *Gnetales* and the anthophyte clade, *Am. J. Bot.*, 1996, 83, 767-780.

Dondua A.K. Developmental Biology. In two volumes. St-Petersburg: Publishing House of St-Petersburg State University; 2005.

Dostal R. Growth correlations in *Bryophyllum* leaves and exogenous growth regulations. *Biol. Plantarum*, 1970, vol. 12, 2, 125-133.

Fagerlind F. Die Samenbildung und die Zytologie bei agamospermischen und sexuellen Artenvon *Elatostema* und einigen nahestetehenden Gattungen nebst Beleuchtung einiger damit zusammenhängender Probleme. *K. Svenska Vet.-Akad.*, 1944, Handl. III, Bd. 21, Heft 4, 1-130.

Gamburg K.Z. Phytohormones and cells. Moscow: Nauka; 1970; 104p.

Gerassimova E.N. Fertilization in *Crepis capillaris*. *Cellule*, 1933, 42(1), 103-148.

Gilbert S.F.; Opitz J.M.; Raff R.A. Resynthesizing evolutionary and developmental biology. *Developmental Biology,* 1996, vol. 173, 2, 357-372.

Gilbert S.F. Development biology. 8-th ed. Sunderland, MA: Sinauer Associates, Inc., 2006, 751 p.

Gordeeva O.F., Mitalipova M.M. Pluripotent stem cells: maintenance of genetic and epigenetic stability and prospects of cell technologies. *Russian Journal of Developmental Biology,* 2008, vol. 39, 6, 325-336.

Gordeeva O.F.; Mitalipova M.M. Pluripotent stem cells: maintenance of genetic and epigenetic stability and prospects of cell technologies. *Ontogenesis,* 2008, vol. 39, 6, 325-336.

Guttenberg H. Grundzüge der Histogenese der höherer Pflanzen. I. Die Angiospermen. Berlin-Nikolassee; 1960.

Guignard J.L., Mestre J.C. L'embryon des Gramineés. *Phytomorphology*, 1971, 20, 2, 190-197.

Isaeva V.V.; Akhmadieva A.V.; Aleksandrova Ya.N., Shukalyuk A.I. Morphofunctional organization of reserve stem cells providing for asexual and sexual reproduction in Invertebrates. *Russian Journal of Developmental Biology,* 2009, vol. 40, 2, 83-96.

Koltsov N.K. Selected works. Moscow: Nauka; 2006, 295 p.

Kordyum E.L. Peculiarities of early ontogenesis of ovule with different type of archesporium in some angiosperms. *Cytology and genetics*. 1968, vol. 2. № 5, 415-428.

Korobova S.N. The movement of sperms in a pollen tube and an embryo sac of angiosperms. In Yakovlev M. S., editor. *Actual Problems of Embryology of Angiosperms*, Nauka, Leningrad, 1979, pp. 5-19.

Kruglova N.N., Batygina T.B., Gorbunova V.Yu., Titova G.E., Seldimirova O.A. Embryological bases of wheat androcliny: Atlas. Moscow: Nauka, 2005, 99 p.

Medvedev S.S. Physiology of plants. St-Petersburg: Publishing House of St-Petersburg State University; 2004, 335p.

Meyen S.V. Basics of paleobotany. Reference manual. Moscow: Nedra; 1987, 403p.

Navashin, S.G. New observations of fertilization in *Fritillaria tenella* and *Lilium martagon, The Diary of X Symp. of Russian naturalists and physicians*, Kiev, 1898, 6, 16-21.

Navashin S.G. On fertilization in Asteraceae and Orchidaceae, *Bull. Imp. l'Acad. Sci. St. Petersburg*, 1900, 13, 335-340.

Navashin, S.G. Results of revision of the fertilization processes in *Lilium martagon* and *Fritillaria tenella*, Selected works, Moscow-Leningrad, Publishing House of AS USSR, 1951, vol.1, 188-192.

Periasamy K. A new approach to the classification of angiosperm embryos. *Proc. Indian Acad. Sci.*, 1977, vol. 9, 1001-1010.

Polevoy V.V. Physiology of plants. Moscow: Vyschaya Shkola, 1989, 464p.

Ray S., Park S.S., Ray A. Pollen tube guidance by the female gametophyte, *Development*, 1997, 124, 2489-2498.

Seydoux G., Braun R.E. Pathways to totipotency: lessons from germ cells. *Cellule*, 2006,127 (5), 891-904.

Schmitz F. Die Blüten-Entwicklung der Piperaceen. Hanstein Bot. Abh. 1872. Bd 2. S. 1-74.

Serebryakova T.I. Bud as the stage of the stem development // Abstract of the VII RBS congress. L. 1983. 232-233.

Serebryakova T.I.; Voronin N.S.; Elenevsky A.G.; Batygina T.B.; Shorina N.I.; Savinykh N.P. Botany with the bases of phytocoenology: anatomy and morphology of plants. Moscow IKC "Academkniga". 2006.

Severtzov, A.N. Morphological regularities of evolution, Moscow-Leningrad, Publishing House of AS USSR, 1939, 601p.

Severtzov A.S. Directionality of evolution. Moscow: Publishing House of Moscow State University; 1990, 270p.

Shamrov I.I. Ovule primordium. In: Batygina T.B., editor. *Embryology of Flowering Plants. Terminology and Concepts. Vol. 1. Generative organs of flower.* St-Petersburg: Mir i Semya; 1994; 132-134.

Shmalgausen I.I. Regulation of morphogenesis in individual development. Moscow, 1964, 136p.

Sinnott E.W. Plant morphogenesis. New York, Toronto, London: McGraw-Hill Book Company, Inc., 1960, 603p.

Smith A. A glossary for stem-cell biology. Edinbourgh UK: Nature, 2006, 441. p. 1060.

Tchaylachyan M.Ch. Regulation of flowering in higher plants. Moscow, 1988, 560p.

Teryokhin E.S., editor. Problems of reproductive biology of seed plants. *Transactions of Komarov Botanical Institute RAS,* vol.8; 1993a; 15-25

Titova G.E. Hypophysis. In: Batygina T.B., editor. *Embryology of Flowering Plants. Terminology and Concepts. Vol. 2. Seed.* St-Petersburg: Mir i Semya; 1997; 352-356.

Tokin B.P. Regeneration and somatic embryogenesis. Leningrad: Leningrad state University Press; 1959.

Tokin B.P. General embryology. Moscow: Vyschaya Shkola, 1987, 480p.

Upadhyay N., Batygina T.B. Embryogenesis and gemmorhizogenesis in *Rauwolfia*. Abstracts of XI International symposium "Embryology and Seed Reproduction", 3-7 July, 1990, Leningrad; 1990; 178.

Van den Berg C., Willemsen V., Hage W. Cell fate in *Arabidopsis* root meristem determined by directional signaling. *Nature*, 1995, vol. 378, 62-65.

Van den Berg C., Weibeek P., Scheres B. Cell fate and cell differentiation status in the *Arabidopsis* root. *Planta*, 1998, vol. 205, 483-491.

Vinogradova G. Polyembryony in Allium ramosum L. and Allium Schoenoprasum L. (Alliaceae) // PhD. Thesis. Komarov Botanical Institute RAS. St. Petersburg 2009. 22p.

Waddington C.H. New patterns in genetics and development. New York, London: Columbia University Press, 1962, 259p.

Winslow T. Stem cells. Scientific progress and future research directions. Bethesda: Nat. Inst. Health, 2001.

Yakovlev M.S. On unity of embryogenesis of angiosperms and gymnosperms. *Transactions BIN AS USSR, ser. VII, 1951*, 2, 231-242.

Glossary

Amphimixis (Gr. *αμφοτέροι* – both, double, *σμίγω* – mix) – is a process of merging of the female and male gametes leading to formation of a new individual. Synonyms: fertilization, sexual process.

Androecium (Gr. *άνδρας* – man, *οἶκος* – house) – is an aggregate of stamens (microsporophylls) in a flower. The number and shape of stamens in a flower varies, which is, first of all, connected with the mode of pollination.

Anther – is a fertile part of the stamen, in microsporangiums of which microsporogenesis takes place, and pollen grains are being formed and maturate.

Antipodals (Gr. *αντί* – anti; *πόδι* – leg) – is a group of cells, the number of which is species-specific and depends on type of embryo sac development, forming the chalazal, rarely chalazal-lateral part of the embryo sac.

Apogamety (Gr. *ἀπό* – without, *γάμος* – gamete, female sexual cell) – is the development of embryoids from cells of egg or antipodal apparatus of the embryo sac.

Apomixis (Gr. *ἀπό* – without, *σμίγω* – mix) – is the formation of the embryo without gametes fusion.

Apospory (Gr. *ἀπό* – without, *σπόρος* – seed) — is the formation of a diploid embryo sac out of the cell of nucellus and integument through mitosis.

Archesporium (Gr. *αρχή* – the beginning, *σπόρος* – spore) – is a cell or a group of cells differentiating into a sporogenous cell or tissue.

Autonomity of the embryo (Gr. *αυτε* – self, *νόμος* – law) – is the property of the embryo characterizing its ability to self regulation and development, independent form the mother plant.

Axis — the notion of some central line, used in engineering, mathematics, physics, biology and other.

Bank of seeds – is the stock of seeds, fruits, parts of fruits, rarely, infructescence. Upon the natural conditions, the main part of the bank of seeds is placed in the soil, the rest at the surface organs of plants. Mature seeds sometimes fall down from shoots immediately, and sometimes remain there for years («bank in the tree crown»). The ability of seeds to remain alive in the ground for a long time is species specific and has formed in the process of formation of each species in course of evolution. Artificial banks of seeds are the storages where the man has created conditions for durable germination ability of seeds.

Brood bud – is a specialized organ of vegetative propagation and dispersal of flowering plants. In the process of brood bud development, its bipolarity is established (adventive roots are laid down) and a new individual is formed. Synonyms: propagule, bulb, tubercle.

Bud (Lat. *gemma*) – is a monopolar structure represented by a reduced shoot or its part. The bud consists of an axis with a growing point at the top (apex), leaves of different age, which cover the axis and each other, and sometimes with primordia of axillary buds, flowers and inflorescence.

Carpel is a leaf-like structure folded in the apocarpous gynoecium conduplicately, i.e. along the main vein, in such a way that the adaxial surface occurs to be inside the cavity, while its edges, more or less tightly, come together, join or grow together.

Central cell of the embryo sac – is a cell named due to the place of its location in the embryo sac. Synonyms: primordium of endosperm, endosperm maternal cell.

Chlorophyll-bearing property of an embryo – is the ability to synthesize chlorophyll during certain period of its development in the seed. **Non-chlorophyllous embryo** – is the absence of the ability to synthesize chlorophyll before seed germination.

Conductive tract of the pistil – is an aggregate of structural elements of the gynoecium enabling its interaction with the male gametophyte in the progamic phase of fertilization, includes all tissues of the pistil (stigma, stile and ovary), which come into contact with pollen grains and growing pollen tubes. Synonyms: conductive tissue.

Correlation — is a statistical interrelation of two or several random values (or values which can be considered random with some acceptable degree of exactness). Changing of one or several of these values lead to a systematic changes of other values. The **correlation coefficient** (positive or negative) serves as a mathematical measure of the correlation of two random values.

Cotyledon (Gr. *κοτυληδών* – cup-shape vessel) – is the first special leaf being laid down in the embryo. Peculiarities of the structure are determined by the place (seed), the time of formation (early stages of ontogenesis), and by the functions (storage, haustorial, protective, photosynthesis).

Critical mass — is the lowest mass of dividing cells necessary for the process of differentiation of separate structures; an inevitable regularity of any pathway of morphogenesis (embryogenesis, embryoidogenesis, organogenesis and histogenesis) passing *in situ*, *in vivo* and *in vitro*.

Cytoplasmatic male sterility (CMS) – is the male sterility appearing as a result of interaction of a specific S-cytoplasm and recessive genes rf rf in homozygotic condition.

Developmental biology – is the science about the driving forces of the development of a living matter studying all phenomena connected with the development of the form and function at different levels of its organization.

Diaspore (Gr. *διά* - through and *σποραδικές* – spreading) – is the unit of the distribution being a part of a plant (or the whole plant) of a different morphological nature, naturally separating from the mother plant. Synonyms: disseminule, propagule.

Diplospory (Gr. *διπλός* - double, *σπόρος* - seed) — is the formation of the embryo sac out of unreduced megasporocyte or diploid megaspore due to abnormal meiosis or its complete replacement with mitosis. Synonym: generative apospory.

Double fertilization – is a combination of one sperm with the egg cell, and the other one (of the same couple) – with the central cell of the embryo sac, forming a diploid embryo and triploid endosperm, for example, upon the Polygonum-type of the embryo sac development.

Doubling structures – are the reserves allowing the system of failures to work at separate stages, and to avoid a complete failure in the system of seed reproduction when there are disorders of the sexual process – meiosis and fertilization.

Ecological embryology – the part of the embryology involving studying of interconnections between factors of the environment, plant behaviour and adaptive peculiarities of the organization of generative and embryonic structures.

Ecological niche – is a position of the species in the ecosystem determined by its relations to the factors of the environment, other species and consumed resources.

Egg cell – is a female gamete out of which, as a result of fertilization, the embryo (new individual) develops.

Embryo – is a primordium of a new plant undergoing all stages of the sporophyte development inside the seed – from zygote to a complex, multi-celled, differentiated for tissues and organs formations.

Embryo sac – is a female gametophyte (formed – cell formation is completed but it is not ready for fertilization, and mature – element specialization is completed, ready for fertilization). There is a variety of embryo sacs (Polygonum, Drusa, Eriostemones, Oenotera, Penaea, etc.), 80% of flowering plants have the Polygonum-type - monosporic, three-mitotic, bipolar.

Embryogenetics – is the discipline which studies mechanisms of genetic determination and regulation of embryological properties, processes and phenomena, genetic consequences, determined by morpho-functional peculiarities of generative and reproductive organs, and structures, possibilities and ways of using cells and structures of generative and reproductive organs for genetic engineering.

Embryogeny (Gr. *ἔμβρυον* – pre-natal fetus, embryo and Fr. *génie* – property, combination of traits in a broad sense of the word; in a narrow sense – the mode of flowering plants reproduction) – is the part of plant embryology describing the development of the sexual embryo.

Embryoid (Gr. *ἔμβρυον* – embryo, *είδος* – species) – is the primordium of the individual forming asexually *in situ*, *in vivo* and *in vitro*.

Embryoidogeny (Gr. *ἔμβρυον* – embryo, *είδος* – species, *γένεσις* – origin) – is a type of an asexual formation of the individual and vegetative propagation of flowering plants *in situ*, *in vivo, in vitro,* with the embryoid being its elementary structure.

Embryology (Gr. *ἔμβρυον* – embryo) – is a science about embryos and embryo development (formal definition). It studies morphogenetic processes at different stages of individual development independently of the fact whether this development is a result of sexual or asexual way of formation.

Endosperm (Gr. *εντός* – inside, *σπέρμα* – seed) – is a heterogeneous tissue developing out of the initial endosperm cell in the embryo sac; the latter appearing during double fertilization as a result of fusion of the central cell of the embryo sac with the sperm.

Epicotyl (Gr. *επί* – on, *κοτυληδών* – cup-shape vessel) – is a part of the stem located under the cotyledon knot.

Epiderm (Gr. *έπί* – on, *δέρμα* – tissue) – is an external primary cover tissue of plants. Synonym: epidermis.

Epiphysis (Gr. *επί* – on, above, *φύσις* – swelling, growing) – is a cell or a group of cells in the apical part of the proembryo being initials of the cortex and epidermis of the shoot.

Evolutionary developmental biology (Eng. – evolution development – Evo-Devo) — is a new trend in developmental biology studying the evolution of ontogenesis and, first of all, researching the evolution of ontogenesis mechanisms.

Failure (reliability theory) — the event which means distortion of an effective state of an object; there is a classification and characteristics of failures.

Fertilization – is a complex multi-stage process involving two main phases: progamic (pollen getting onto the stigma) and postgamic (double fertilization).

Fruit (Gr. *φρούτο*) – is a morphological formation developing out of the ovary after fertilization of seed ovules.

Generative cell (Gr. *γεννώ* – to produce, to give birth) – is a smaller cell of a two-cellular pollen grain appearing as a result of a differentiating mitosis and, at the following division, forming two sperms. Synonyms: antheridial, spermatogenous, spermiogenous cell.

Genetic heterogeneity of seeds – is the ability of plants to form, in the same seed, sexual embryos and embryoids of different origin (sexual and asexual) and of various genetics (uniparental or biparental).

Gynoecium (Gr. *γυνή* – woman, *οἶκος* – house) – is an aggregate of carpels (megasporophylls) in an angiosperm flower. In a flower there are one or several carpels which can be free, i.e. not growing together with each other (apocarpous gynoecium – Gr. *ἀπό* – apart and *καρπός* – fruit) or growing together in a whole unity (syncarpous gynoecium – Gr. *σύν* – together).

Heterocarpy (Gr. *ἕτερο-* – another, different, *καρπός* – fruit) – is a genetically determined property of the species to form fruits of different morphological and anatomic structure on the same individual.

Heterospermy (Gr. *ἕτερο-* – another, different, *σπέρμα* – seed) — is the availability of seeds, at one plant, differing in size, weight, color, morphology, anatomic structure, genetic characteristic, biochemical composition, character of germination and other properties. Synonyms: heterogeneity, dissimilarity, non-consistency, different quality of seeds.

Hierarchy (Anc. Gr. *ἱεραρχία*, from *ἱερός* «sacred» and *ἀρχή* «governing») — is the order of subordination of lower elements to higher ones, their organization into a structure like a tree; the principle of governing in centralized structures.

Homeostasis (Anc. Gr. *ὁμοιοστάσις* from *ὅμοιος* – same, similar, and *στάσις* – standing, immobility) – is self-regulation, ability of the open system to maintain consistency of its inner state due to coordinated reactions aimed at keeping the dynamic balance.

Hypophysis (Gr. *ὕπο* – under, below, *φύσις* – swelling, growing) – is a cell or a group of cells in the basal part of the proembryo which give rise to elements of the embryonic root.

Hypostase – (Gr. *ὑπο-* – under, below and *στάση* – stop) – is a tissue differentiated at the base of the nucellus and integuments.

Integument (Fr. *in* – no and *tégument* – coat) – is the structure of the ovule surrounding megasporangium (=nucellus).

Living form – is the appearance, or habitus of a plant, involving its surface and underground organs and existing in harmony with the environmental conditions. Synonyms: biomorph (Gr. *βίος* – life, *μορφή* – form).

Megasporocyte – is a cell of sporophyte transforming into a megaspore after meiosis and giving rise to a female gametophyte (embryo sac). Synonym: megaspore mother cell.

Megasporogenesis – is a process in the course of which megasporocytes undergo meiosis giving birth to megaspores. There are three types of megasporogenesis: monosporic, bisporic and tetrasporic.

Meiosis (Gr. *μείωση* – decreasing) — is a special type of cell division when the whole number of chromosomes is decreased two times. Out of each cell with a diploid number of chromosomes, four haploid sexual cells are formed in the course of meiosis. The ability to decrease the number of chromosomes two times comprises the biological essence of meiosis and guarantees maintaining the same diploid number of chromosomes at the sequence of generations.

Micropyle (Gr. *μικρός* – small, *πύλη* – gate, entrance) – is the channel formed by integument (or integuments) closed at the nucellus top and enabling penetrating of the pollen tube into the embryo sac at porogamy.

Microsporangium (Gr. *μικρός* – small, *σπόρος* – seed, *ἀγγείον* – container) – is a sporogenous part of the anther. The development of microspores and pollen grains take place in microsporangia. The synonyms: anther lobe, pollen sac.

Microsporangium wall – is the layers of microsporangium cells (parietal, endothecium, medium layer, tapetum), surrounding the sporogenous tissue.

Microspore (Gr. *μικρός* – small, *σπόρος* – spore, seed) – is the first cell of the gametophyte generation being formed in the microsporangium as a result of two meiotic divisions of microsporocyte.

Microsporocyte (Gr. *μικρός* + *σπόρος* + *κύτός* – vessel, container) – is a sporogenous cell entering meiosis after completing mitosis in it. Synonyms: microspore mother cell, meiocyte.

Microsporogenesis (Gr. *μικρός* + *σπόρος* + *γένεσις* – origin, appearing) – is the process of microspores formation through meiotic division of microsporocytes developing in anther lobes.

Mycorhiza – is the evolutionarily formed and structurally arranged symbiosis, necessary for one or both partners, between a fungi (adapted to living in the ground and inside the plant) and the root (or other organ being in contact with a substrate and providing transport of substances) of a living plant. Mycorhiza develops in specialized organs of plants where a close contact between symbiotes is a result of a simultaneous development of a plant and fungi. Mycorhizal symbioses represent the continuum from mutual to parasite relationships, and the position in it is defined by symbiotes' species, environment factors and the age stage of the symbiosis.

Nucellus (Lat. *nucellus*) – is the main structure of the ovule where sporogenous cells and female gametophyte are being laid down and developed. Synonyms: megasporangium, macrosporangium.

Ovule – is an organ of a seed plant formed at the placenta and being a place of megasporogenesis, female gametophyte formation and fertilization process, and which, after complicated transformations, gives rise to a seed. The ovule is a transformed megasporangium of seed plants, being protected by sterile coats.

Parthenogenesis (Gr. *παρθένος* – virgin, *γένεσις* – origin) – is the development of the embryo without fertilization. In plants, parthenogenesis is divided into reduced and unreduced ones. In the first case, the embryo has haploid number of chromosomes, while in the other case, – diploid or polyploid number (as a result of apomeiosis).

Particulation (Lat. *particula* – part, particle) – is a longitudinal partitioning of a plant, mainly its underground organs (caudex, vertical rootstock, main root, stem-root tubers), into separate living parts (particles) capable of independent existence and development.

Phragmoplast (Gr. *φραγμα* – partition, *πλαστικός* – molded) – is a cytoskeleton component of plants appearing at the stage of a cell passing on to division.

Phytocoenosis (Gr. *φυτεύω* – plant, *κοινός* – comon) — is a part of plants characterized by relative consistency of structure, floristic composition and conditions of the abiotic environment. Phytocoenosis is one of the main objects of geobotany (phytocoenology).

Pistil (Lat. *pistillum*), carpel (Lat. *carpellum*) — female reproductive organ of flower plants out of which the fruit is formed. It consists of ovary, in which cavity there can be found ovules, and stigmas sometimes uplifted by a stile (stilodium).

Placenta (Gr. *πλακούντας* – flat cake) – is a place of laying down and attaching of an ovule to the carpel. Placenta is quite a big outgrowth which sometimes almost fills the cavity in the ovary.

Polar nuclei (Gr. *πολικός* – axe, pole) – are nuclei of the central cell of the embryo sac (endosperm maternal cell).

Polarity (Ancient Greek *πόλος* – axis of rotation) — the definite direction, in a figurative sense, - contradiction of two senses.

Pollen grain (Gr. *γύρη* – pollen and Lat. *granum* – grain) – is a multicellular structure formed due to one or two mitotic divisions in the microspore, as a result of which a male gametophyte appears. Synonym: male gametophyte.

Pollen tube – is an element of the male gametophyte enabling transportation of sperms to the embryo sac for further double fertilization.

Polygonum-type of embryo sac development – is a monosporic, three-mitotic, bipolar. This type of the embryo sac is the most widespread among angiosperm (more than 80% of species). Normally, being formed out of a chalasal cell of the microspore tetrad. A formed embryo sac, bipolar, seven-celled, eight-nuclear.

Population (Gr. *πολλοί* – many) – is an aggregate of individuals of the species within which a relatively free exchange of genes takes place, resulting in individuals of the same population being genetically quite similar. This is a minimal self-reproducing group of individuals of the same species having a quite big amount, and inhabiting the same area for evolutionarily long span of time and, for a period of several generation being isolated from similar groups.

Potential seed productivity — is a maximum possible amount of seeds which is capable of producing a plant, population or phytocoenosis at a certain period of time on the condition that all ovules laid down in flowers are able to form mature seeds.

Proembryo (Gr. *προτού* – before, pro- and *ἔμβρυον* – embryo) – is an embryo at early stages of development from the zygote to embryoderm differentiation.

Propagation – in a broad sense of the word (*sensu lato*) includes all the processes which lead to increase in number of biological units, with old and new structures having succession.

Protocorm (Gr. *πρώτος* – initial, first, *κορμός* – stem, shooting, tubercule) – is a tuber-shaped germ in orchids and some Lycopodiophyta. Synonyms: embryonic tubercle, germ.

Protosome (Gr. *πρώτος* – first, initial, *σομα* – body) – is a general name of germs of highly specialized parasite flower plants. Protosome is a non-differentiated vegetative body (some) developing out of root or stem poles of embryo and carrying out the function of search and exploitation of **host-plant or fungi symbiont.**

Real (factual) seed productivity — is the amount of capable seeds produced by a plant to give birth to one individual.

Renewal – is the reproduction and increase of the amount of the population.

Reproduction – is the formation of one new individual (through sexual or asexual process).

Reproductive biology – is a field of research studying the processes of propagation and reproductive cycles at different levels of hierarchy (organism, population-species and biocoenotic levels). Reproductive biology reflects the dependence of propagation on environmental conditions.

Reproductive effort — is the part of material and energy resources focused at the process of reproduction.

Reproductive success of plants – is a number of gene copies transmitted to the next generation which is also capable of propagation. The parameter characterizing the living condition and final achievements in the struggle for life. The notion has different meaning depending on the level of organization of biosystems – at the **individual level** it is connected with fertility of a plant, at the **population level** – it depends on age, sexual content and living state of individuals. Reproductive success is also determined by structural and hormonal readiness of seeds for germination.

Reserve (Lat. *reservare* — save, preserve) – is the store of something in case of demand, the source of new facilities and strengths.

Sarmentation (Lat. *sarmentum* — vein, offspring) — is a formation of offsprings out of top, lateral and adventive buds on stems, rootstock, roots, which after settling get separated from the mother plant.

Seed – is one of the elements of reproduction, propagation and dispersal typical to seed plants; contains the primordium or primordia (embryos of different origins) of new plant and specialized storage tissue (endosperm, perisperm, etc.) placed into the seed coat (sporoderm).

Seed dormancy – is a naturally appeared in the course of evolution state allowing to survive conditions unfavourable for germination. Types of dormancy are the provoked and organic one.

Seed propagation – is an increase of the amount of individuals of the same species through seeds.

Sperm (Gr. *σπέρμα* – seed) – is the male sexual cell, or male gamete. Sperm is formed as a result of division of the generative cell either already in the pollen grain or, after its germination, in the pollen tube. Sperms contain the haploid set of chromosomes and are characterized by considerably transformed nuclear-cytoplasmatic relations.

Sporogenous tissue – is a tissue being formed as a result of periclinal divisions of archesporial cells.

Stamen – a generative organ of a flower being a microsporophyll transformed in the course of the evolution (Gr. *мicrosporophyllum*). The stamen consists of a proximal sterile part – stamen filament and a distal fertile part – anther.

Stemness – is an integrity of all properties characteristic to stem cells. Mainly, this is an ability to self-renewal, differentiation and specialization.

Strategy (Anc. Gr. *στρατηγία* – the art of the chief commander) – is a general, undetailed plan of some activity embracing a long period of time, the way to achieve a complicated goal through solving intermediate tactic objectives.

Strategy of plants' life – is survival of an organism and a reproduction consequently replacing each other in ontogenesis. **Reproduction strategy** is accordingly associated with K- and r-strategy which plants realize passing on through their life cycles.

Suspensor (Lat. *suspendere* – to suspend) – is a specialized organ of the embryo, which is differentiated out of the basal cell of the proembryo (*cb*) or its derivatives, and serves for providing the developing embryo with nutritious and growing substances, and often, for its moving deeper into the endosperm.

Symbiosis – is a joint existence of organisms belonging to different species and not similar with each other. Symbiosis can be arranged on the basis of trophic or spatial relations and differ in balance of benefit and damage: antagonistic symbiosis – parasitism, beneficial only to one of the partners – commensalism, having mutual benefit – mutualism. The main properties of symbiosis lie in the duration and specificity of relations between the partners, appearing of new properties, not characterized for separate organisms, formation of specific structures, morphological co-adaptation of symbiotes and regulation of relations between them at the genetic and physiological level.

Symmetry (Anc.Gr. *συμμετριαι* — compatibility) in biology — is a regular location of similar (alike) parts or forms of a living organism, aggregate of living organisms in relation to the centre or the symmetry axis (radial or bilateral). **Asymmetry** (Gr. *α-* — without and *συμμετριαι* — compatibility) — absence of symmetry; used to describe organisms initially deprived of symmetry. **Dissymmetry –** is the secondary loss of symmetry or its separate elements in the course of evolution.

Synergid (Gr. *σύν* – together, *έργο* – work) – is a cell-satellite of the egg cell. These cells (usually there two of them) together with the egg cell form the unified complex in the embryo sac – egg apparatus.

Syngamy (Gr. *σύν* – together, *γάμος* – marriage) – is the process of fusion of two haploid gametes (egg cells and sperms), which results in forming of the zygote.

System of reliability – is an ability of an organism to liquidate or transform damages appearing spontaneously or induced by extreme impact, so that they would not threaten the organism's survival or its reproduction.

Tapetum (Gr. *τάπης* – paving layer, carpet) – is a multifunctional tissue which is directly attached to sporogenous tissue of the anther, enabling meiosis, normal development of micrspores and maturation of pollen grains.

Totipotency (Lat. *totus* – the whole, integral, *potentia* – power, strength, ability) — is the capability of a cell (or cells) with all morphogenetic abilities (i.e. potential) typical for given individual and realized through various pathways of morphogenesis.

Triple fusion – is the process of fusion of polar nuclei of the central cell with the sperm's nucleus which results in the formation of the primary endosperm nucleus.

Vegetative cell (Lat. *vegere* – to excite, grow) – is a larger cell of the pollen grain producing pollen tube during its germination.

Vegetative propagation – is an increase in the amount of individuals of the species through buds, shoots, roots, embryoids.

Viviparity (Lat. *vivus* – alive, *pario* – to give birth) – is a way of reproduction and propagation with generative diaspore, containing the embryo, or vegetative diaspore, without the period of dormancy, forming germs (propagules) already on the mother organism.

Zygote (Gr. *ζευγν-* – combined together) – is a cell being formed as a result of fusion of the female gamete (egg cell) with the male one (sperm).

Index

A

B

C

D

E

F

G

Q

R

S

T

U

V

X

Z